Necessary DoD Range Capabilities to Ensure Operational Superiority of U.S. Defense Systems Testing for the Future Fight

Committee on Assessing the
Physical and Technical Suitability of
DoD Test and Evaluation Ranges and Infrastructure

Board on Army Research and Development

Division on Engineering and Physical Sciences

A Consensus Study Report of

The National Academies of
SCIENCES • ENGINEERING • MEDICINE

THE NATIONAL ACADEMIES PRESS
Washington, DC
www.nap.edu

THE NATIONAL ACADEMIES PRESS 500 Fifth Street, NW Washington, DC 20001

This activity was supported by Contract W911NF-18-D-0002 with the Office of the Secretary of Defense. Any opinions, findings, conclusions, or recommendations expressed in this publication do not necessarily reflect the views of any organization or agency that provided support for the project.

International Standard Book Number-13: 978-0-309-49857-9
International Standard Book Number-10: 309-49857-0
Digital Object Identifier: https://doi.org/10.17226/26181

Limited copies of this report may be available through the Board on Army Research and Development, 500 Fifth Street, NW, Washington, DC 20001; (202) 334-3111.

Additional copies of this publication are available from the National Academies Press, 500 Fifth Street, NW, Keck 360, Washington, DC 20001; (800) 624-6242 or (202) 334-3313; http://www.nap.edu.

Copyright 2021 by the National Academy of Sciences. All rights reserved.

Printed in the United States of America

Suggested citation: National Academies of Sciences, Engineering, and Medicine. 2021. *Necessary DoD Range Capabilities to Ensure Operational Superiority of U.S. Defense Systems: Testing for the Future Fight*. Washington, DC: The National Academies Press. https://doi.org/10.17226/26181.

The National Academies of
SCIENCES • ENGINEERING • MEDICINE

The **National Academy of Sciences** was established in 1863 by an Act of Congress, signed by President Lincoln, as a private, nongovernmental institution to advise the nation on issues related to science and technology. Members are elected by their peers for outstanding contributions to research. Dr. Marcia McNutt is president.

The **National Academy of Engineering** was established in 1964 under the charter of the National Academy of Sciences to bring the practices of engineering to advising the nation. Members are elected by their peers for extraordinary contributions to engineering. Dr. John L. Anderson is president.

The **National Academy of Medicine** (formerly the Institute of Medicine) was established in 1970 under the charter of the National Academy of Sciences to advise the nation on medical and health issues. Members are elected by their peers for distinguished contributions to medicine and health. Dr. Victor J. Dzau is president.

The three Academies work together as the **National Academies of Sciences, Engineering, and Medicine** to provide independent, objective analysis and advice to the nation and conduct other activities to solve complex problems and inform public policy decisions. The National Academies also encourage education and research, recognize outstanding contributions to knowledge, and increase public understanding in matters of science, engineering, and medicine.

Learn more about the National Academies of Sciences, Engineering, and Medicine at www.nationalacademies.org.

The National Academies of
SCIENCES • ENGINEERING • MEDICINE

Consensus Study Reports published by the National Academies of Sciences, Engineering, and Medicine document the evidence-based consensus on the study's statement of task by an authoring committee of experts. Reports typically include findings, conclusions, and recommendations based on information gathered by the committee and the committee's deliberations. Each report has been subjected to a rigorous and independent peer-review process and it represents the position of the National Academies on the statement of task.

Proceedings published by the National Academies of Sciences, Engineering, and Medicine chronicle the presentations and discussions at a workshop, symposium, or other event convened by the National Academies. The statements and opinions contained in proceedings are those of the participants and are not endorsed by other participants, the planning committee, or the National Academies.

For information about other products and activities of the National Academies, please visit www.nationalacademies.org/about/whatwedo.

COMMITTEE ON ASSESSING THE PHYSICAL AND TECHNICAL SUITABILITY OF DoD TEST AND EVALUATION RANGES AND INFRASTRUCTURE

DANA "KEOKI" JACKSON, NAE,[1] MITRE Corporation, *Chair*
DARRYL AHNER, Air Force Institute of Technology
KAREN BUTLER-PURRY, Texas A&M University
GRAHAM V. CANDLER, University of Minnesota
GORDON FORNELL, United States Air Force, Retired
DERRICK HINTON, Scientific Research Corporation
ROB KEWLEY, stimlytics.cloud, LLC
LAURA J. McGILL, Sandia National Laboratories
HANS MILLER, MITRE Corporation
HEIDI C. PERRY, Massachusetts Institute of Technology Lincoln Laboratory
GARY POLANSKY, Sandia National Laboratories
KARL F. SCHNEIDER, Department of the Army, Retired
WILLIAM WILSON, Carnegie Mellon University

Staff

LIDA BENINSON, Senior Program Officer, Board on Higher Education and Workforce, *Study Director*
WILLIAM "BRUNO" MILLONIG, Director, Board on Army Research and Development (BOARD)
STEVEN DARBES, Program Officer, BOARD
CHRIS JONES, Senior Finance Business Partner, BOARD
CAMERON MALCOM, Research Assistant, BOARD
CLEMENT MULOCK, Program Assistant, BOARD
RYAN MURPHY, Program Officer, Air Force Studies Board
LINDA WALKER, Program Coordinator, Board on Physics and Astronomy
SAMUEL ZINKGRAF, Research Assistant, BOARD (through May 2021)

Consultant

ROBERT POOL, Writer

NOTE: See Appendix D, Disclosure of Unavoidable Conflicts of Interest.
[1] Member, National Academy of Engineering.

BOARD ON ARMY RESEARCH AND DEVELOPMENT

KATHARINA McFARLAND, U.S. Army (retired), *Chair*
MICHAEL BEAR, BAE Systems, *Vice Chair*
ANDREW ALLEYNE, University of Illinois, Urbana-Champaign
DAVID AUCSMITH, University of Washington
JAMES BAGIAN, NAE[1]/NAM,[2] University of Michigan
JOAN BIENVENUE, University of Tennessee
LYNN DUGLE, Independent Consultant
JOHN FARR, United States Military Academy at West Point
GEORGE "RUSTY" GRAY III, NAE, Los Alamos National Laboratory
WILLIAM HIX, U.S. Army (retired)
GREGORY JOHNSON, Lockheed Martin
DUNCAN McGILL, Mercyhurst University
CHRISTINA MURATA, Deloitte
ADITYA PADHA, Deloitte
ALBERT SCIARRETTA, CNS Technologies, Inc.
GEOFFREY THOME, SAIC
JAMES THOMSEN, Seaborne Defense, LLC
JOSEP TORRELLAS, University of Illinois, Urbana-Champaign

Staff

WILLIAM "BRUNO" MILLONIG, Director
STEVEN DARBES, Program Officer
SARAH JUCKETT, Program Officer
TINA LATIMER, Program Coordinator
CAMERON MALCOM, Research Assistant
CLEMENT MULOCK, Program Assistant
CHRIS JONES, Senior Finance Business Partner

---

[1] Member, National Academy of Engineering.
[2] Member, National Academy of Medicine.

Preface

Our nation's warfighters go into combat to fight and win equipped with weapon systems that must operate under the harshest conditions, against determined and capable adversaries. They rightfully expect that these weapons have been tested and proven effective under operationally relevant conditions, against realistic threats that represent the battlefield they will confront. The Department of Defense's (DoD's) test and training range enterprise makes possible this essential developmental and operational testing, and these key resources for national security rest on the dedicated contributions of thousands of military personnel, civil servants, defense contractors, and representatives of national laboratories and federally funded research and development centers. They are at the heart of the range enterprise, and labor under extremely challenging conditions, generally unseen and unknown to the public due to the criticality of their work. The future viability of DoD's range enterprise depends on addressing dramatic changes in technology, rapid advances in adversary military capabilities, and the evolving approach the United States will take to closing kill chains in a Joint All Domain Operations environment. This recognition led DoD's Director of Operational Test and Evaluation (OT&E), the Honorable Robert Behler, to request that the National Academies of Sciences, Engineering, and Medicine examine the physical and technical suitability of DoD's ranges and infrastructure through 2035.

The study committee brought a diverse set of perspectives and expertise to the questions posed in the statement of task, with members from industry, academic, and government backgrounds, versed in the application of emerging technology, the operational use and test of advanced weapon systems, the rapidly changing landscape of digital technologies, and the organizational and budgetary complexity faced by the OT&E community and the range enterprise. The committee readily acknowledges that the extraordinary diversity of DoD missions and test environments, and the large quantity of range locations and installations, precluded an exhaustive evaluation of all range capabilities and gaps in relation to the future OT&E landscape. Nevertheless, the committee is confident that the findings and conclusions described in this report represent common themes fully supported by a survey of several of the most significant ranges, and an extensive review of prior studies and reports on OT&E needs and the implications for the range enterprise. The committee also notes that this unclassified study addresses certain key challenges and solutions at a general level due to the sensitive nature of many U.S. military capabilities and the intelligence gathered on current and future threats posed by U.S. adversaries. The combined background in national security matters of the committee underpins its belief that this report's recommendations address DoD's overarching range enterprise needs, while recognizing that the second, classified phase of this study will provide important additional detail and context regarding the test and evaluation requirements for the ranges posed by new weapons capabilities and threat characteristics.

The committee is grateful for the contributions of a wide range of noted experts and thought leaders in military weapon systems development, test, and evaluation; innovation and emerging technologies; software-intensive systems and digital capabilities; and the operational challenges both current and future faced by the U.S. military. Likewise, we received outstanding support from representatives of many test and training ranges spanning warfighting domains across land, sea, air, space, and cyberspace, who contributed their time and insights Many of the experts who participated in the study's workshops and committee meetings have a distinguished record of public service, including in the military, and we thank them for that service to our nation. We also are pleased to acknowledge the gracious support from Mr. Robert Arnold, Senior Advisor of Sustainable Ranges, and Dr. Raymond O'Toole, acting Director, Operational Test and Evaluation, in providing connections and access to key officials, DoD resources, and reference materials that were indispensable to the study committee. It has been a privilege to work with these dedicated public servants and subject matter experts on this important priority for the nation's defense.

> Keoki Jackson, *Chair*
> Committee on Assessing the Physical
> and Technical Suitability of DoD Test and
> Evaluation Ranges and Infrastructure

Acknowledgments

The committee would like to thank the following individuals for providing input to this study:

ANDRE' "DRE" ABADIE, U.S. Army Futures Command
JAMES AMATO, Army Test and Evaluation Command
ZACH BARBER, Nevada Test and Training Range
LISA BARNEBY, Point Mugu Sea Range
STEPHEN BEARD, Missile Defense Agency
ROBERT BEHLER, Former Director, Operational Test and Evaluation
MARC BERNSTEIN, Assistant Secretary of the Air Force
    (Acquisition, Technology, and Logistics)
ASHTON BURKE, Test Resource Management Center
DEVIN CATE, U.S. Air Force
ERIC CLINTON, Test Resource Management Center
VICTORIA COLEMAN, U.S. Air Force
CHRIS COLLINS, Under Secretary of Defense for Research and
    Engineering
RYAN "RHINO" CONNER, Electromagnetic Spectrum Superiority,
    U.S. Air Force
MICHAEL CONTRATTO, 96th Test Wing
JAMES COOKE, U.S. Army
DENNIS CRALL, Joint Staff J6
FREDERICK CRAWFORD, Institute for Defense Analyses
MISSY CUMMINGS, Duke University

BILL DARDEN, Atlantic Test Range
EVAN DERTIEN, Air Force Materiel Command
JESSIE DIETZ, Pacific Multi-Domain Training Experimentation Capability
FRED DRUMMOND, Office of the Secretary of Defense
JASON ECKBERG, U.S. Air Force
VIV EDWARDS, Nevada Test and Training Range
JOHN ELLIS, Missile Defense Agency
FRED ENGLE, Office of the Assistant Secretary of Defense for Readiness
ERIC FELT, Air Force Research Laboratory
JOHN FIORE, Naval Surface Warfare Center
MATT FUNK, NAVAIR Acquisition and Tech Support Division
JOHN GARSTKA, Office of the Secretary of Defense
JEFFREY GERAGHTY, Wright-Patterson Air Force Base
CONRAD GRANT, Johns Hopkins University Applied Physics Laboratory
WILLIAM GREENWALT, American Enterprise Institute
DEREK GREER, NAVAIR Integrated Battlespace Simulation and Test
ED GREER, Formerly with the Office of Developmental Test & Evaluation
ROBERT GRIMES, Nevada Test and Training Range
SCOTT HOSCHAR, Atlantic Test Range
ARTHUR HUBER, Air Force Materiel Command
CHRIS JARBOE, Atlantic Test Range
PAUL KAWSHNAK, Aberdeen Proving Ground
PAUL KETRICK, National Cyber Range Complex
MICHAEL LABER, Point Mugu Sea Range
EDGAR LACY, Naval Surface Warfare Center Dahlgren Division
BRIAN LEONG, Pacific Multi-Domain Training Experimentation Capability
PETER LEVINE, Institute for Defense Analysis
RYAN "CHEECH" LUCERO, Nevada Test and Training Range
MIKE MACKINAW, Pacific Multi-Domain Training Experimentation Capability
JOSHUA MARCUSE, Google
DONALD MARTIN, Nevada Test and Training Range
BARRY MOHLE, Naval Surface Warfare Center Dahlgren Division
CARL MURPHY, Test Resource Management Center
BRIAN NOWOTNY, Test Resource Management Center
JOHN OKUMA, Institute for Defense Analyses
DANIEL OSBURN, 412th Test Wing
RAYMOND O'TOOLE, Director, Operational Test and Evaluation

BRENT PARKER, Pacific Multi-Domain Training Experimentation Capability
DAN PATT, Thomas H Lee Partners
JOHN PEARSON, Office of the Secretary of Defense Air Warfare
JANE PINELIS, Joint Artificial Intelligence Center
CARROLL "RICK" QUADE, Deputy Assistant Secretary of the Navy Research, Development, Test & Evaluation
JACK RILEY, Pacific Multi-Domain Training Experimentation Capability
STEVE ROGERS, Air Force Space Command
LEE ROSEN, SpaceX
DANIEL ROSS, Naval Surface Warfare Center Dahlgren Division
ROBIE SAMANTA ROY, Lockheed Martin
GEORGE RUMFORD, Test Resource Management Center
DAVID SAYRE, Missile Defense Agency
SCOTT SBUKOFF, Pacific Multi-Domain Training Experimentation Capability
HERMAN "HEMET" SCHIRG, Nevada Test and Training Range
CAPT WILLIAM SELK, Commanding Officer, VX-1
KENNETH SENECHAL, NAVAIR
ARUN SERAPHIN, Senate Armed Services Committee
JASON STEWART, Atlantic Test Range
JACOB SUGGS, Missile Defense Agency
ROBERT TAMBURELLO, Test Resources Management Center
MICHAEL TAYLOR, SpaceX
NEIL THURGOOD, Office of the Assistant Secretary of the Army
BRYAN TITUS, Air Force Space Command
GIL TORRES, Naval Air Warfare Center Weapons Division
RODNEY TRAYLOR, Nevada Test and Training Range
ANDREW TREE, Point Mugu Sea Range
DAVID TREMPER, Office of the Secretary of Defense
EDWARD TUCKER, Arnold Engineering Development Complex
ROBERT VARGO, Atlantic Test Range
JEFFREY WHITE, Secretary of the Army
MICHAEL WHITE, Under Secretary of Defense for Research and Engineering
KEVIN WILLIAMS, Missile Defense Agency
LEMUEL WILLIAMS, Missile Defense Agency
GEOFFREY WILSON, Test Resource Management Center
ERIC "GLOCK" WRIGHT, Nevada Test and Training Range
GREG ZACHARIAS, Chief Scientist, Director of Operational Test and Evaluation
PETER "ZUPP" ZUPPAS, Nevada Test and Training Range

The committee would also like to express its gratitude to Maya Thomas and Christopher Lao-Scott, Research Librarians at the National Academies Research Center, for their assistance with fact checking.

Acknowledgment of Reviewers

This Consensus Study Report was reviewed in draft form by individuals chosen for their diverse perspectives and technical expertise. The purpose of this independent review is to provide candid and critical comments that will assist the National Academies of Sciences, Engineering, and Medicine in making each published report as sound as possible and to ensure that it meets the institutional standards for quality, objectivity, evidence, and responsiveness to the study charge. The review comments and draft manuscript remain confidential to protect the integrity of the deliberative process.

We thank the following individuals for their review of this report:

Sharon Beerman-Curtin, Strategic Consulting, LLC,
Russel Caflisch, NAS, New York University,
Stephen Di Domenico, Coldsquared Consulting,
Kathleen Dussault, Lemon Grove Associates,
James Michael Gilmore, Institute for Defense Analysis,
Lester Lyles, NAE, Independent Consultant,
Chris Maston, Georgia Tech Research Institute, and
Julie Ryan, Wyndrose Technical Group.

Although the reviewers listed above provided many constructive comments and suggestions, they were not asked to endorse the conclusions or recommendations of this report nor did they see the final draft before its release. The review of this report was overseen by John Tracy,

NAE, Boeing (retired). He was responsible for making certain that an independent examination of this report was carried out in accordance with the standards of the National Academies and that all review comments were carefully considered. Responsibility for the final content rests entirely with the authoring committee and the National Academies.

Contents

EXECUTIVE SUMMARY   1

1   INTRODUCTION   8
    Study Charge, 10
    Military Ranges Past, Present, and Future, 13
    Fundamental Themes, 15
    Five Categories of Solutions, 19
    Structure of the Report, 25
    References, 26

2   AN ENVISIONED FUTURE OF OPERATIONAL TEST AND
    EVALUATION   27
    The Future of Warfighting, 28
    The Envisioned Future of Military Test Ranges, 32
    Enabling the Envisioned Future of Military Ranges, 38
    References, 41

3   TESTING FOR FUTURE COMBAT: MULTI-DOMAIN
    OPERATIONS, CONNECTED CONCURRENT KILL CHAINS,
    AND MITIGATING ENCROACHMENT   43
    Testing for the Multi-Domain Battlespace, 44
    A Joint Program Office to Support DoD Multi-Domain Testing
        Needs, 54
    Mitigating Encroachment to Support Future Combat Testing, 57
    References, 66

4 DIGITAL INFRASTRUCTURE NEEDS FOR OPERATIONAL
   TESTING   68
   Modeling and Simulation, 69
   Increasing the Usability and Value of Data, 80
   References, 90

5 SPEED-TO-FIELD: RESTRUCTURING THE
   REQUIREMENTS AND RESOURCES PROCESSES FOR
   DoD TEST RANGES   92
   Program Requirements Drive Range Funding Investments, 93
   Colors of Money for Range Modernization and Maintenance, 94
   Strategies to Improve Test Range Modernization, 98
   References, 103

6 CONCLUSION AND SUMMARY OF RECOMMENDATIONS
   BY ACTOR   104
   The Recommendations—By Stakeholder, 105

APPENDIXES

A Statement of Task and Completion Matrix   111
B Site Visit Summaries   114
C Committee Member Biographies   122
D Disclosure of Unavoidable Conflicts of Interest   129
E Abbreviations and Acronyms   131

Executive Summary

Rigorous operational testing (OT) of weapon systems procured by the U.S. Department of Defense (DoD) is fundamental to ensuring that these sophisticated systems not only meet their stated requirements, but also perform under realistic operational conditions when faced by determined adversaries employing their own highly capable offensive and defensive weaponry. Without adequate OT, operational commanders would be unable to make the most effective use of their capability and warfighters would lack confidence in the weapons they bring to the fight or, worse, may inadvertently put themselves in harm's way because they do not have a fundamental understanding of their weapons' capabilities and limitations. DoD's test and training range enterprise provides the geography, infrastructure, technology, expertise, processes, and management that make safe, secure, and comprehensive OT possible. However, the range enterprise, along with the talented and committed range workforce that makes the system function, is under great stress. Unless prompt action is taken to address both longstanding and emerging challenges, including test capacity, modernization, digital infrastructure, encroachment, and resources, DoD's ranges will be unable to support timely or adequate OT in the future.

The challenges facing the nation's range infrastructure are both increasing and accelerating. Limited test capacity in physical resources and workforce, the age of test infrastructure, the capability to test advanced technologies, and encroachment impact the ability to inform system performance, integrated system performance, and the overall pace of testing. Investments in the U.S. test infrastructure and changes in test and evaluation (T&E) methodologies and handling of data are necessary to inform the delivery of lethal, survivable, reliable, and affordable weapon systems to the field at a speed that is relevant to the operational need. This study draws on testimony from senior military officers and officials from operational, acquisition, and test backgrounds as well as on test and training experts, leading technologists, leaders from relevant commercial enterprises, and individuals with deep experience in DoD and congressional budget processes. The study committee conducted virtual and physical site visits to a representative sample of test ranges; collected test range inputs on modernization, sustainment, operations, and resource challenges; and reviewed prior studies and reports from the office of the Director of Operational Test and Evaluation (DOT&E), the military service test organizations, and the Test Resource Management Center (TRMC). This report makes a set of interdependent recommendations that the committee believes will put the DoD range enterprise on a modernization trajectory to meet the needs of OT in the years ahead.

The report emphasizes the following three fundamental themes:

1. **Future combat will demand connected kill chains in a Joint All-Domain Operations (JADO) environment.** It is critical that DoD architects, specifies, develops, and tests systems to ensure that they are highly effective when fielded in this new reality. DoD acquisition processes, organizational stovepipes, test methods, and range infrastructure that were optimized for the testing of individual weapon systems in single domains will be inadequate to test future integrated weapon systems in the way that they will be operated in machine-speed warfare that crosses all combat domains, including land, sea, air, space, and cyberspace.
2. **Digital technologies are dramatically reshaping the nature, practice, and infrastructure of test.** The weapon systems of today and tomorrow are fundamentally enabled by data and software, and DoD test ranges will be no different. The rapidly increasing importance of autonomy, artificial intelligence (AI), and machine learning across defense systems is creating novel challenges for OT. Furthermore, the advent of digital twins and high-performance modeling and simulation (M&S) is enabling new ways of testing, even as combinations of new domains and operational constraints increasingly make virtual testing the only practical approach for certain applications.
3. **Speed-to-field is today's measure of operational relevance, which is in turn a continuously moving target.** Enabled by the global proliferation of many digital, software, and communications-based technologies, U.S. adversaries are rapidly and continuously

EXECUTIVE SUMMARY 3

deploying new generations of weapons designed to negate U.S. warfighting advantages. At the same time, new weapon systems are employing never-fielded technologies, which are also evolving at a pace enabled by Moore's Law. Usable weapon systems are fielded promptly, but there is a need for continuous testing and assessment.

To address the challenges tied to these themes, the committee developed conclusions and recommendations that fall into the following five broad categories:

1. **Develop the "range of the future" to test complete kill chains in JADO environments.** It is essential that the range enterprise accommodate new concepts of operation and new test approaches for realistic operational testing, which includes enabling infrastructure for system-of-systems integrated testing and the interoperability of multiple ranges across diverse domains. [Recommendation 3-1]
2. **Restructure the range capability requirements process for continuous modernization and sustainment.** Enabling speed-to-field while maintaining the rigor of operational test and evaluation will require rapid range modernization for new weapons technologies and new threats. At the same time, key capabilities need to be sustained and even augmented to ensure required test capacity and throughput, while mitigating the issues caused by encroachment both in the physical and radio frequency environments. [Recommendations 3-2, 3-3, 3-4, 3-5]
3. **Bootstrap a new range operating system for ubiquitous M&S throughout the weapon system development and test life cycle.** Many of today's DoD programs cannot be tested effectively in live testing alone. High-fidelity virtual testing can improve readiness and the likelihood of success for actual hardware testing and may be the only environment to do certain types of tests. However, widespread and standardized use of M&S for operational test will depend on a new M&S infrastructure, significant cultural changes within the test community, and new approaches to the validation of M&S in an ever-changing threat and technology environment. [Recommendation 4-1]
4. **Create the "TestDevOps" digital infrastructure for future operational test and seamless range enterprise interoperability.** Redefine the enterprise-supported core digital standards and capabilities for TRMC and test ranges to take advantage of DoD's scale for software, data, networks, AI, cybersecurity, and M&S.

Make model-based engineering, the unbroken digital thread, and continuous integration/continuous delivery software practices the foundation for range agility, rapid test evolution, and speed-to-field. Ultra-high-bandwidth information flows must become frictionless, on-demand, and secure. [Recommendations 4-2, 4-3]

5. **Reinvent the range enterprise funding model for responsiveness, effectiveness, and flexibility.** The resource needs of today and tomorrow reflect the reality of rapidly changing technology and threats; sustained capital investment for creation, upgrade, and maintenance of long-life range systems; and increasing demands for cross-domain system-of-systems testing with seamlessly integrated M&S. Including DOT&E earlier and continuously in the requirements development and acquisition processes will better establish and certify the timeliness and adequacy of range investments. [Recommendations 5-1, 5-2; Conclusions 5-1, 5-2]

Table ES.1 maps the report recommendations and key conclusions to the themes laid out above.

TABLE ES.1 Report Recommendations and Key Conclusions Mapped to Themes

Theme	Recommendations and Key Conclusions
Develop the "range of the future" to test complete kill chains in JADO environments	Conclusion 3-1: The lack of a Department of Defense or joint publication set of definitions for multi-domain operations and cyber-physical systems can result in different operational use cases.
	Conclusion 3-2: Testing ranges are not optimized for testing end-to-end kill chains; they were not designed for collaborations with other ranges, and they lack the framework and infrastructure to test concurrent and connected kill chains.
	Recommendation 3-1: To enable a range of the future that is capable of testing kill chains and multi-domain operations that can integrate effects across National Defense Strategy modernization areas, the Secretary of Defense should address the need to enable Department of Defense ranges to provide regular venues to "test as we fight" for acquisition and prototyping programs in a joint multi-domain battlespace of integrated systems.

TABLE ES.1 Continued

Theme	Recommendations and Key Conclusions
Restructure the range capability requirements process for continuous modernization and sustainment	Conclusion 3-3: Encroachment leads to the inability to demonstrate mission capability and identify deficiencies due to lack of access to the physical and electromagnetic spectrum space with which to conduct test and evaluation. This creates operational risk as DoD will have to field weapon systems that have not been tested against certain threats. Recommendation 3-2: To ensure the ability to validate the survivability of Department of Defense (DoD) weapon systems against a realistic operational threat environment across air, sea, land, space, and cyberspace domains, DoD should identify and prioritize bands that cover U.S. military operational and test requirements which should be protected from sell-off to preserve these capabilities. Recommendation 3-3: The Test Resource Management Center (TRMC) should assess current and projected commercial radio frequency communications technologies and spectrum allocations for secure, agile, high-bandwidth operational test needs. In addition, TRMC should determine the feasibility of developing new large-scale enclosed testing facilities combined with expanded modeling and simulation to support electromagnetic spectrum activities not suitable for open-air testing. Recommendation 3-4: The Department of Defense should broaden the authority of the Test Resource Management Center to address issues of internal encroachment by reviewing internal range policies and actions to ensure that the test groups retain adequate mission space and prevent the placement of equipment or infrastructure that could potentially interfere with test operations. The Director of Defense Research and Engineering for Advanced Capabilities should be granted the authority to mitigate disputes arising over internal encroachment concerns and provided additional funding to manage internal encroachment. Recommendation 3-5: The Test Resource Management Center should develop a strategy that assesses the use of and potential investment in suitable allied resources for open-air testing. This strategy should include criteria for the usage of allied resources and areas of potential investment to include range space available, data collection, security risks, and support facilities.

continued

TABLE ES.1 Continued

Theme	Recommendations and Key Conclusions
Bootstrap a new range operating system for ubiquitous M&S throughout the weapon system development and test life cycle	Recommendation 4-1: A Department of Defense joint program office should establish a shared, accessible, and secure modeling and simulation (M&S) and data ecosystem to drive development and testing across the life cycles of multiple supporting programs. M&S should be planned from early concept development to support the entire life cycle of the system, from requirements generation, through design development, integration and test, and sustainment. Uncertainty quantification should be employed to identify the primary sources of uncertainty in the understanding of the system being developed and to define an integrated testing and simulation activity to reduce those uncertainties to an acceptable level.
Create the "TestDevOps" digital infrastructure for future operational test and seamless range enterprise interoperability	Recommendation 4-2: A Department of Defense joint program office should adopt and promulgate modern approaches for standardization, architectural design, and security efforts to address data interoperability, sharing, and transmission challenges posed by the complexity of next-generation systems. The joint mission office should determine how to develop and maintain a protected data analysis tool and model repository for testing, increase the interconnectivity of test ranges, and ensure the development of data protocols for the real-time transfer of data at multiple classification levels.
	Recommendation 4-3: The Test Resource Management Center should continue monitoring and supporting the Assured Development and Operation of Autonomous Systems (ADAS) Project, and prioritize efforts to develop a common set of standards, measurement approaches, and operational scenarios from which to evaluate the performance of artificial intelligence (AI) and autonomous systems, while recognizing that testing approaches may differ between AI and autonomous systems.

TABLE ES.1 Continued

Theme	Recommendations and Key Conclusions
Reinvent the range enterprise funding model for responsiveness, effectiveness, and flexibility	Recommendation 5-1: The Joint Requirements Oversight Council (JROC) should consult regularly with the Director of Operational Test and Evaluation (who is an advisor to the JROC) about the test requirements for systems considered by the JROC. This consultation should include an evaluation of current testing capabilities, facilities shortfalls, and plans to address these shortfalls. Recommendation 5-2: The Office of the Secretary of Defense should either allow an exemption or set shallower expenditure benchmarks for the first 2 years of test modernization programs. This will reflect realistic expense curves for the technologies and projects needed to test next-generation programs and complex integration. Conclusion 5-1: New mechanisms and funding limits for applying minor military construction are necessary for responsive test and evaluation activities. Conclusion 5-2: There exists a need for the Department of Defense to pilot new process and authorities for funding ranges and infrastructure to make them simpler, more responsive, and more effective.

1

Introduction

To protect itself from attacks by foreign forces, the United States relies on its armed services, which in turn rely on weapons and other systems to provide the tools needed to successfully neutralize adversary combat capabilities. Maintaining the armed services' warfighting advantage requires a steady stream of new and improved weapons and technologies. A crucial step in acquiring and using these assets is testing their effectiveness and suitability on Department of Defense (DoD) ranges. DoD has testing ranges that span the globe where new military technologies are tested based on real threats, tasks, and environments to ensure their combat readiness. These ranges are a vital aspect of the nation's defense, but will they be able to adequately test the increasingly complex military technologies of the future, at the pace required?

Former Director of DoD Operational Test and Evaluation (OT&E) Robert Behler has noted that the U.S. test range system dates to the years during and after World War II, with the most significant updates having been carried out during the Cold War.[1] The committee recognizes that there have been further upgrades and modernization in the 30 years following the end of the Cold War. However, Raymond O'Toole, the current acting director of OT&E, asserted at the committee's January 2021 public workshop that the ranges have not kept pace with testing demands, technology development, or the capabilities of adversaries

---

[1] From remarks delivered at December 4, 2020, committee meeting; recording available at https://www.nationalacademies.org/our-work/assessing-the-physical-and-technical-suitability-of-dod-test-and-evaluation-ranges-and-infrastructure.

(NASEM, 2021, p. 2). Consider just some of the major environmental shifts that have affected weapon system development and testing since that time period:

- The United States now faces not one, but two peer military adversaries in China and Russia, and China is an economic powerhouse that has broadened the realm of great power competition far beyond avenues of military conflict.
- Other adversaries such as North Korea and Iran have rapidly developed missile, cyber, and nuclear weapon technology that can now threaten the U.S. homeland. Furthermore, rival nations have made a concerted effort to outpace the United States in weapons technologies of the future, including artificial intelligence, hypersonics, and space systems.
- Commercially available internet technology, mobile communications, cloud computing, and ubiquitous software have fundamentally reshaped the architecture and testing of weapon and support systems. Moreover, commercial demand and development are creating robust competition for the physical geography and radio frequency spectrum that OT&E has relied upon for decades.

The stewardship and use of DoD test ranges relies on multiple stakeholders to effectively test the nation's defense systems. The developmental test and evaluation (DT&E) community is responsible for funding and executing the upgrades on which these future capabilities depend and therefore must be fully included in policy, resourcing, and allocation discussions moving forward. It is only through the full participation of all the stakeholders that lasting change can be realized. Urgent and substantial changes to the modernization, sustainment, operation and resourcing of the range enterprise are required to support the scale and diversity of weapon system testing and to meet the challenges posed by rapid insertion of new technology over the next 10–15 years.

The consequences of inaction will be severe, and recovery will be difficult. Technologies such as artificial intelligence, hypersonics, cyber weapons, and directed energy are creating new test capability requirements for DoD ranges. Rapidly improving threats, particularly from peer adversaries such as China and Russia, make the need to test a system's survivability just as important as testing its lethality. Peer and non-peer adversaries, including actors like North Korea, increasingly employ asymmetric capabilities such as cyberattacks. These trends, when extrapolated to 2035, demand a new approach to modernizing DoD's ranges' technical and physical attributes. This new approach must not only preserve the current core capabilities, but also take a more holistic look at the aggregate capabilities needed to address the software-intensive nature of future weapon systems, as opposed to the too-often piecemealed upgrades that occur today.

STUDY CHARGE

Against this backdrop, DoD's Office of the Director, OT&E requested the Board on Army Research and Development of the National Academies of Sciences, Engineering, and Medicine to perform a study assessing the physical and technical suitability of DoD test and evaluation ranges, infrastructure, and tools for determining the operational effectiveness, suitability, survivability, and lethality of military systems (see Box 1.1, Statement of Task). While the ranges' staffing and organizational structures are clearly crucial to their success, the statement of task specified that the study should focus not on the personnel and staffing but on the facilities themselves.

This study is the first of two studies that were requested by DoD's Director of OT&E. It is an unclassified review that was designed as a stand-alone study but that will also provide the foundational elements for a follow-on classified study that is scheduled to begin before the official publication of this report. The purposes of conducting this study at the unclassified level were to make it possible to include on the committee as wide a range as possible of members from the science and engineering community as well as to build maximum awareness of the serious challenges to be addressed today. The follow-on study, which will be informed by the contents of this report, will assess how well the ranges are able to simulate the threats, threat countermeasures, and capabilities of near-peer adversaries and to test DoD systems in future operational scenarios. The goal of carrying out the two studies is that, by taking advantage of the full range of the nation's science and technology community in the unclassified study while also having access to complete information on adversaries' capabilities in the classified study, the two together will offer a comprehensive assessment of DoD's testing and evaluation ranges and infrastructure.

The Committee's Approach

To carry out the statement of task and evaluate the nation's military ranges, the National Academies Board on Army Research and Development (BOARD) assembled a study committee composed of experts from the military, industry, academia, and government. The committee, assisted by BOARD staff members, assembled a broad collection of written and graphic information related to the ranges and OT&E, including many

BOX 1.1
Statement of Task

The National Academies of Sciences, Engineering, and Medicine will convene an ad hoc committee to assess the physical and technical suitability of the Department of Defense's (DoD's) ranges, infrastructures, and tools used for test and evaluation (T&E) of military systems' operational effectiveness, suitability, survivability, and lethality across all domains (land, sea, air, space, and cyberspace). Specifically, the committee will:

1. Assess the aggregate physical suitability of DoD's ranges to include their testing capacity, the condition of their infrastructure, security measures, and encroachment challenges.
2. Assess the technical suitability of ranges to include spectrum management, instrumentation, cyber and analytics tools, and their modeling and simulation capacity.
3. Evaluate the following attributes for each range:
    o Physical Attributes of Range: Do ranges allow for full exercise of tested systems in the manner they will be used to achieve their mission?
    o Electromagnetic Attributes of Range: Can the system under test, and emulated threats to the system, access and utilize spectrum as designed and needed?
    o Range Infrastructure: Can range instrumentation properly and fully assess system performance and record test data (as well as training data that could be applied to T&E requirements)? Can range tools adequately process and transmit test data and efficiently provide test results?
    o Test Infrastructure Security: How secure are ranges, infrastructure, and test capabilities against physical and cyber intrusion that could lead to exploitation of weapon systems performance data by an adversary?
    o Encroachment Threats and Impacts: What are the existing and potential future encroachment threats and impacts (physical space, spectrum, alternative/competing DoD uses)?
4. The committee will recommend how the DoD can address and/or mitigate any existing or anticipated deficiencies, and test and evaluate future technologies anticipated to arrive between now and 2035, including discussion of planning and resource allocation for the overall test range enterprise. These technologies include, but are not limited to:
    o Directed energy, hypersonic systems, autonomous systems, artificial intelligence, space systems and threats, 6th generation aircraft, advanced acoustic and non-acoustic technologies for undersea warfare, and advanced active electronic warfare/cyber capabilities.

previous reports authored by various components of DoD, the National Academies, the Congressional Research Service, RAND Corporation, and other groups. This literature formed the foundation on which the committee based its judgments. It was supplemented by presentations, typically held via Zoom, by multiple military officials, both active and retired, as well as other experts familiar with military ranges and the challenges of OT&E in the current environment.

On January 28–29, 2021, the committee held a public virtual workshop, Assessing the Physical and Technical Suitability of DoD Test and Evaluation Ranges and Infrastructure. Over the course of 2 days, the committee heard presentations from representatives of the Test Resource Management Center, the individual services' testing and evaluation departments, various other service groups involved in the development and testing of military systems, national laboratories, universities, and industry. In April 2021, *Key Challenges for Effective Testing and Evaluation Across Department of Defense Ranges: Proceedings of a Workshop—in Brief* was published to summarize the workshop's presentations and what the committee learned from them (NASEM, 2021).

Over the course of the study, committee members and BOARD staff carried out seven site visits, some in person and some virtually, at a diverse selection of ranges that were as representative as possible of the wide array of challenges facing OT&E over the coming decade and a half. A summary of the site visits is provided in Appendix B. During those visits the committee and staff heard from a wide range of military representatives with intimate knowledge of the day-to-day workings of the ranges and OT&E, and the information gleaned from these visits, combined with the knowledge and insights that the individual committee members brought to the process, formed much of the foundations for the deliberations that resulted in this report.

For those deliberations the committee members met regularly, both in full committee and in subcommittees, from December 2020 through July 2021. Depending on their expertise and interest, different committee members contributed to different parts of this report, but all writing, from the narrative to the findings and recommendations, was reviewed by and agreed on by the entire committee. This report is the result of that process.

The extraordinary diversity of DoD missions and test environments, and the large quantity of range locations and installations, precluded an exhaustive evaluation of all range capabilities and gaps in relation to the future OT&E landscape. DoD's test and training ranges number over 500 in total, including the 23 major facilities in the Major Range and Test Facility Base (MRTFB). Additionally, DoD does not currently have standardized and comprehensive reporting on test ranges and facilities.

INTRODUCTION 13

To assess the current physical and technical state of the test ranges, the committee selected representative ranges spanning all domains (land, sea, air, space, and cyberspace) to provide insights on the aggregate challenges with operational testing unique to each domain. This strategy enabled the committee to report on concerns and conditions that were articulated by multiple ranges, services, and agencies. The committee further recognizes that each of DoD's test ranges will face specific challenges and opportunities unique to the individual facility or organization that are not addressed in this report.

This unclassified study addresses certain key challenges and solutions at a general level due to the sensitive nature of many U.S. military capabilities and the intelligence gathered on current and future threats posed by U.S. adversaries. Other topics in the statement of task are not readily addressed without referencing controlled unclassified information (CUI). The second, classified phase of this study will provide important additional detail and context regarding the test and evaluation requirements for the ranges posed by new weapons capabilities and threat characteristics. Appendix A includes a matrix mapping the committee's work against the statement of task (SOT), with the disposition of task areas not addressed or partially addressed in this report.

MILITARY RANGES PAST, PRESENT, AND FUTURE

DoD operates a large number of ranges, spanning all of the services, which are used to test and evaluate the effectiveness of military systems and train operators in every domain: land, sea, air, space, and cyberspace. These ranges and their infrastructure and associated tools (and personnel) are a critical component of the DoD acquisition community and its systems development process, and they play critical research, experimentation, development, test, and training roles in the never-ending modernization efforts aimed at ensuring that the country's warfighters are provided with the operational superiority its citizens expect if they are to deal effectively with the nation's adversaries. Among the technologies that must be capable of being tested at the nation's ranges are directed energy weapons, hypersonic platforms, autonomous systems, artificial intelligence, space systems, 6th generation aircraft, long-range munitions, acoustic and non-acoustic technologies for undersea warfare, advanced electronic warfare/cyber capabilities, chemical and biological defense, and hard and buried target countermeasures.

The performance and credibility of military weapon systems against threats and adversaries are foundational to U.S. deterrent capability and battlefield advantage. While this fact has not changed, the fundamental nature of warfare has shifted as a result of the information revolution, the emergence of linked battle networks, and changes to the concepts of operation resulting from what has been termed previously a "revolution in military affairs" (Mowthorpe, 2005; Murray, 1997). The concept of networked warfare—linking sensors, command and control, and precision weapons across platforms—became visible to the world in the first Gulf War. Adversaries, including China and Russia, observed and reacted, developing their own integrated battle networks that can hold U.S. forces at risk.

Future U.S. deterrent and combat capability will depend on the ability to close the kill chains and dismantle adversary kill chains. The fight will span all domains of conflict—air, land, sea, space, and cyberspace—and extend to competition before the active phase of conflict. However, performance and credibility against emerging threats can only be demonstrated through the testing of production-representative systems in realistic operational conditions, against realistic representations of adversarial capabilities. Here, speed is critical for all activities that lead to the fielding of combat capability—including test and evaluation. As adversaries rapidly develop and deploy their own advanced weapon systems in an iterative fashion, it is critical that the range enterprise and infrastructure support testing of the latest systems and technologies in order to keep up with and stay ahead of the most current—and anticipated—threats. The variety of adversary weapons in development, and the speed with which they are being tested and deployed, is at a pace and scale that exceeds anything the United States has seen in a generation or more. The U.S. Air Force Chief of Staff, General Charles Q. Brown, states this bluntly in the title of his August 2020 report "We Must Accelerate Change or Lose (ACOL)" (Brown, 2020).

Today's information technologies and digital infrastructure create a fundamentally new dynamic for the practice of OT&E, and the tools, approaches, infrastructure, and skill sets must keep pace. As Marc Andreesen commented in 2011, "Software is eating the world" (Andreesen, 2011). His essay described how software-centric products and services were taking over large segments of the economy while fundamentally disrupting value chains across the physical world. Just as no industry, system, or product is immune, OT&E and DoD's range infrastructure must address the interrelated sets of challenges and opportunities. A major theme of the Defense Innovation Board's report on DoD software acquisition is that "software is different than hardware"—software intensive systems are the core of U.S. offensive and defensive fighting capabilities, and software is never "done" (Defense Innovation Board, 2019). In some cases, the science and approaches behind the testing of new software-driven technologies such as artificial intelligence (AI) and machine learning are still being developed. While these facts are daunting, the same set of information and digital technologies, when applied appropriately and at scale, will provide the solution to the challenges if the necessary steps are taken to modernize DoD's range infrastructure and test methods.

FUNDAMENTAL THEMES

Given this situation and in view of the testimony of experts from across DoD, technology development, and commercial enterprise combined with data collected from site visits, the committee structured the report around the following three fundamental themes, which are further expanded in the paragraphs below:

1. Future combat will demand connected kill chains in a joint all-domain operations environment.
2. Digital technologies are dramatically reshaping the nature, practice, and infrastructure of testing.
3. Speed-to-field is today's measure of operational relevance, which is in turn a continuously moving target.

Future Combat Will Demand Connected Kill Chains in a Joint All-Domain Operations Environment

As described in the book *The Kill Chain*, platforms and weapons are the tools of the military, but ultimately "the ability to prevail in war, and thereby prevent it, comes down to one thing: the kill chain. . . . It involves three steps: The first is gaining understanding about what is happening. The second is deciding about what to do. And the third is taking action that creates an effect to achieve an objective" (Brose, 2020, p. xviii). This concept is not new, and it is even documented as a "mission thread with a kinetic outcome" in the DoD Mission Engineering Guide (DoD, 2020, p. 36). However, the networked and interconnected nature of today's kill chains requires more from the DoD test enterprise. At the committee's January 2021 workshop, Col. Jason Eckberg, DoD's deputy director of electromagnetic spectrum dominance, describes the shift required from one-on-one tests that focus on a platform's lethality and survivability to "tests with multiple components that assess overall force effectiveness." (NASEM, 2021, p. 7)

As stated in the 2018 National Defense Strategy, "We face an ever more lethal and disruptive battlefield, combined across domains, and conducted at increasing speed and reach—from close combat, throughout overseas theaters, and reaching to our homeland" (DoD, 2018, p. 3). In order to test for the future fight, with composable kill chains across domains, platforms, networks, and command and control systems, the integration between different platforms and systems will be increasingly tested. Furthermore, test approaches and range capabilities will have to be as agile and adaptable as future weapon systems, as those systems and threats evolve and thus change warfighting tactics, techniques, and procedures. Because kill chains will span many or all warfighting domains, from undersea to space and everything in between, test approaches and test ranges will require the ability to stitch together multiple ranges alongside virtual and constructive models of both "blue" and "red" forces. As Marc Bernstein, the chief scientist for the office overseeing all Air Force acquisition, said at the committee's January 2021 workshop, it will be necessary to combine many, if not all, of the nation's test ranges into a very complex "range of ranges" (NASEM, 2021, p. 6).

Digital Technologies Are Dramatically Reshaping the Nature, Practice, and Infrastructure of Test

The weapon systems of today and of the future are defined as much by software as hardware, as are the adversary threats U.S. forces face. Battle networks are central to current and future kill chains, and information technology is at the heart of cyber and electronic warfare. However, as noted at the committee's January 2021 workshop by David Tremper, director of electronic warfare in the Office of the Secretary of Defense, the ranges lack the software-defined agile threat systems that would allow testing against more representative threats (NASEM, 2021, p. 5). Artificial intelligence and machine learning create novel challenges that require the development of underlying test science to address learning systems that adapt and respond to their environments and the systems deployed against them. Bernstein warns that it will be AI against AI, and test and evaluation (T&E) must model those threats (NASEM, 2021, p. 5). Moreover, as noted by Joshua Marcuse, head of strategy and innovation at Google and formerly executive director of the Defense Innovation Board, some military ranges seem to have barely entered the digital age at all, while facing challenges in exponentially increasing requirements for data and the accompanying instrumentation, collection, telemetry, communication, storage, processing and analytics (NASEM, 2021, p. 9).

Some of the same technologies also create opportunities to reinvent the test ranges for this digital world. Digital engineering techniques and tools offer the promise of co-developing test systems in parallel with the weapon systems they will support and rapidly updating test capabilities to keep pace with the evolution of weapons and the threats they face. "TestDevOps" approaches can mirror the "DevSecOps" agile development processes and platforms increasingly used in system development, enabling comparable responsiveness through automation and continuous integration/continuous delivery. The expanded integration of modeling and simulation (M&S) with real-world testing in live–virtual–constructive environments will enable the creation of cutting-edge test environments simulating realistic threat densities as well as the adaptability of threat systems. M&S will also support new ways of testing integrated kill chains and enable the evaluation of holistic unit actions and the training of extended forces on how to use new weapon systems. Other M&S benefits include the replication of threats and capabilities that are too sensitive or too dangerous to be reproduced in the real world and the potential to carry out far more tests over wider sets of conditions than are practical on physical test ranges. For all these benefits, digital capabilities create their own unique challenges, such as cybersecurity, the need for rigorous model validation, and limitations in the ability to adequately replicate real-world uncertainty that may constrain applicability for AI systems. Moreover, these digital environments for testing must be both timely and continually refreshed. As described in the Government Accountability Office (GAO) report on the F-35 program's Joint Simulation Environment (JSE), technical problems with the simulator have put necessary test capabilities years behind schedule, delaying completion of OT&E and the next production milestone decision (GAO, 2021). This example demonstrates how critical digital infrastructure has become in proving operational suitability and the impact to U.S. forces if that infrastructure does not meet the required pace. It also illustrates the many technical and programmatic challenges inherent in development of the complex, high-fidelity, validated M&S environments required for the testing of the most advanced weapon systems that must be understood and addressed to realize the full promise of M&S for test and training.

Speed-to-Field Is Today's Measure of Operational Relevance, Which Is in Turn a Continuously Moving Target

The vice chairman of the Joint Chiefs of Staff, General John Hyten, stated that "inserting speed into everything the Defense Department does is a priority." The reason: "when you look at our competitors, large and small, one of the things that you find that they have in common is they're moving very, very fast. And we are not" (Cronk, 2020). While testing is only a part of the sequence of events in fielding a weapon system, it is generally the step that allows a declaration of operating capability or the approval for rate production. Neither the operational test system nor the ranges that support testing are optimized to increase speed of capabilities to the field. As Behler noted at the January 2021 workshop, "You could update the F-35 plane as fast as an iPhone app . . . and you wouldn't actually be any faster relative to your adversary because you would still need a year for me to test it" (NASEM, 2021, p. 2). Pacing functions include readiness of the systems under test and the required test infrastructure, which drive the ability and schedule of those systems to undergo and pass operational testing. Our ability to replicate current operational threats is likewise painfully slow. At the public workshop, Ed Greer, the former deputy assistant secretary of defense for developmental test and evaluation, shared how it takes an average of 3 to 5 years from the time that intelligence is collected on threats to the time those threats are instantiated into testing, during which time adversaries can build new systems faster than intelligence centers can build models.[2] As a result of this, test ranges and programs must anticipate "pop up" testing requirements driven by new tactics and techniques, or emerging threats, will be the norm in the future, rather than the exception.

There are many factors that slow DoD's ability to test the latest systems against the latest threats, and often its range enterprise creates limiting factors. Some of these bottlenecks are related to the capacity of the ranges, driven by dated, limited, or unreliable sensing and communications systems; the constrained availability of unique test facilities; the need to link multiple ranges to support a test; and limitations on the ability to conduct tests simultaneously or around the clock. Others, as noted above, tie to the ability to update software-intensive test support systems at the pace required or information security systems and processes that prevent rapid switching between security levels. At the January 2021 workshop, Marcuse noted that despite the digital revolution of the past several decades, testing remains optimized for hardware (NASEM, 2021, p. 4), and does not take advantage of tools, processes, and automation that enable speed and responsiveness in test infrastructure, test execution, and test problem remediation.

Perhaps the greatest challenge to speed and responsiveness for the test ranges, however, is the multi-year process by which funding is allocated to sustain, operate, and modernize the range enterprise. Funds come from a multiplicity of centralized and distributed sources, with different rules for the application and timing of expenditures, which cannot be practically combined or redistributed to meet the most pressing priorities, and often tie the hands of the teams charged to ensure the readiness of the ranges. Beyond the inefficiency and uncertainty of the budget process, in many cases the range investments are inadequate to the meet the capability needs for the weapon systems on the schedules they must support. As shared by Behler at the committee's December 2020 meeting,

---

[2] From remarks delivered on January 29, 2021, at the public workshop; recording available at https://www.nationalacademies.org/event/01-28-2021/assessing-the-physical-and-technical-suitability-of-dod-test-and-evaluation-ranges-and-infrastructure-meeting-2-and-workshop.

a rule of thumb in the test community is that approximately 1 percent of acquisition spending should be allocated to test and evaluation infrastructure. For critical domains, such as space, the planned investment levels fall far below that level.[3]

FIVE CATEGORIES OF SOLUTIONS

The study committee identified five sets of solutions that reflect the actions required to address the key cross-cutting themes. Within these categories there are complementary individual recommendations for addressing the challenges and opportunities described in the chapters of this report. Taken together, these committee judgments will enable DoD to develop, implement, and sustain the range enterprise capabilities needed to meet the challenges posed by the three major themes (connected kill chains, digital technology, and speed-to-field).

Across the three main themes the committee's judgments fall into 5 categories of solution sets:

1. Develop the "range of the future" to test complete kill chains in joint all-domain environments.
2. Restructure the range capability requirements and process for continuous modernization and sustainment.
3. Bootstrap a new range operating system for ubiquitous M&S throughout the weapon system development and test life cycle.
4. Create the "TestDevOps" digital infrastructure for future operational test and seamless range enterprise interoperability.
5. Reinvent the range enterprise funding model for responsiveness, effectiveness, and flexibility.

Develop the "Range of the Future" to Test Complete Kill Chains in JADO Environments

In his opening remarks to the committee in December 2020, Behler emphasized that the ranges must be able to integrate systems and domains to enable the promise of combined arms for decisively closing blue force kill chains in the future fight.[4] However, OT&E and the range enterprise have focused on the test of single programs and systems against their individual operational requirements, and collaborative effects have not

---

[3] From remarks delivered at December 4, 2020, committee meeting; recording available at https://www.nationalacademies.org/our-work/assessing-the-physical-and-technical-suitability-of-dod-test-and-evaluation-ranges-and-infrastructure.

[4] Ibid.

typically informed major test requirements. Large-scale exercises, such as Black Dart, Emerald Flag, and Orange Flag, have been conducted in recent years to investigate the feasibility of seamless operations across services and domains, and these exercises also illuminated the challenges in performing rigorous and repeatable operational testing with this scale and complexity. With the new centrality of integration and all-domain warfighting in the national defense strategy, the infrastructure of the range of the future must instantiate the ability to bring together systems-of-systems across warfighting domains, including land, air, sea, space, and cyberspace, and to measure the effectiveness of end-to-end kill chains performing against threats across those domains.

The range enterprise must be able to connect the ranges together with speed and agility and to perform efficient and effective command and control of tests across this distributed range-of-ranges while maintaining a high level of safety and security. This means that all performing and supporting organizations must have a clear and common understanding of multi-domain operations concepts and definitions and of the requirements for effective testing of today's diverse cyber-physical systems. Connections via secure, high-bandwidth lines of communication governed by common data standards, processes, and procedures will enable the ranges to collect, share, store, manage, and analyze the massive volumes of test data. A new organizational construct embodied in a joint program office with supporting policy and doctrine changes is recommended to manage the framework for testing of kill chains across systems and technologies, starting from use cases and concepts of operation, continuing through capability development and evolution, enabling integrated kill chain testing, and ultimately providing feedback to both the operational and acquisition communities for informed operations and future developments.

Growing encroachment also poses particular threats for integrated kill chain testing, whether in the physical, radiofrequency, or cyber domains, as tests increasingly span geographical regions and make use of extensive spectrum resources for sensing, communications, and weapons effects. The committee explores the potential mitigations for loss of space or spectrum, with recommendations on managing internal encroachment within DoD's span of control as well as on managing external encroachment via U.S. government action or technological solutions and workarounds.

Restructure the Range Requirements and Resourcing Process for Continuous Modernization and Sustainment

Range modernization requirements are primarily determined by program test requirements, which are established by programs in the acquisition phase. While establishment of test requirements is intended to occur via test and evaluation master plans (TEMPs) developed early in the acquisition process, often the understanding of test approaches and resulting range needs is immature. Focus and priority for the ranges and Test Resource Management Center (TRMC) is on test requirements over the next 3 to 5 years, so active preparation for the technologies, techniques, and range infrastructure needed in the next 10 to 15 years receives less attention and resources. Furthermore, test requirements are difficult to modify once testing needs and priorities are established, limiting the flexibility and agility of the range enterprise as the ranges are faced with addressing rapidly evolving weapons technologies and continuously advancing threats. The committee finds that recapitalization and modernization for broader or longer-term use beyond individual test program requirements is not incentivized. Greater attention to range enterprise modernization needs early in the acquisition process by the Joint Requirements Oversight Council (JROC), including tracking of current and projected range capability gaps, is recommended to address the observed range requirements shortfalls.

In addition to range modernization requirements driven by individual program test needs, the ranges must now react to a new set of capability and resource challenges and gaps driven by kill chain testing of integrated systems in representative Joint All-Domain Operations (JADO) environments. The current piecemeal, program-driven requirements process results in many projects to develop individual range capabilities, while structures, accountability, and processes to link and integrate these capabilities are immature or ad-hoc. A new joint program office is recommended to develop, maintain, and update cross-service and cross-domain mission threads, JADO test approaches and an integrated systems test requirements framework, and resulting range and infrastructure demands to test for the future fight.

Encroachment on the nation's test ranges, including both physical intrusions and limitations, and reduced access to electromagnetic spectrum resources, is also driving test range requirements for the future. Actions are recommended to increase DoD's oversight and prevention of internal range encroachment actions, while strengthening DoD's role in U.S. government in regulation and allocation of adjoining geographical regions and military-relevant radio frequency (RF) bands. Additional steps to identify potential Allied open-air range resources, and determine how emerging commercial communications capabilities can be best used to address high-bandwidth range requirements, will also help to mitigate encroachment concerns for the future.

Bootstrap a New Range Operating System for Ubiquitous M&S Throughout the Weapon System Development and Test Life Cycle

M&S are rapidly increasing in importance for test and evaluation of weapon systems. Realistic physical replication of the quantity and diversity of adversary threats that will be faced by U.S. systems in combat is becoming impractical on U.S. test ranges. Often, testing of sensitive capabilities in open-air venues is inadvisable due to increasing vulnerability to adversary monitoring and surveillance. Moreover, the logistical complexity and cost for extensive physical testing of integrated systems across full operational envelopes for multiple use cases is prohibitive.

M&S provide necessary and useful capabilities that can address many of these challenges. Digital engineering and model-based engineering approaches may provide an integrated virtual representation of weapon systems, their interfaces, the operational environment, and diverse threats across the full system life cycle from requirements development and architecture through design, manufacturing, integration, test and evaluation, and sustainment. Increasing availability and reduced costs of high-performance computing and validated functional and physics-based models, coupled with "big data" analytic capabilities, and the application of machine learning expand the opportunity to complement physical testing with large numbers of virtual tests conducted in parallel rather than sequentially. Statistical techniques can be applied to optimize test approaches combining physical and virtual testing, increasing the value of each physical test while minimizing expensive and time-consuming range testing and potentially reducing the quantity and variety of required test assets. Modern, agile, and iterative software development paradigms, such as DevSecOps (for Development-Security-Operations) can increase the pace of test development and execution in the virtual environment, allowing more rapid adaptation to changing operational concepts and threat scenarios.

However, M&S are not a "silver bullet." Fundamental challenges with integration, physical system equivalence, validation, and realistic development schedules and costs remain to be addressed (Wolfe, 2021). While reducing the projected increase of demands on the range enterprise, and offering approaches that may be more cost-effective and practical than physical testing alone, it is unlikely that M&S will reduce the overall amount of testing on the ranges. At a minimum, testing will be required to validate models, quantify uncertainties, and understand the limits of simulations. To achieve the greatest benefits from M&S for the T&E and the ranges, resources must be applied to create a centralized, persistent and accessible M&S environment and library of validated models that can be shared across the T&E enterprise, coupled with education and training to achieve effective utilization. M&S must be applied from the earliest phases of concept development, and sustained throughout the life cycle of the programs and systems addressed. Robust and efficient multi-level security approaches must be developed to manage diverse classification guidance and protect sensitive information while moving at the pace required to stay ahead of adversary threats.

Create the "TestDevOps" Digital Infrastructure for Future Operational Test and Seamless Range Enterprise Interoperability

DevSecOps—the combination of agile development methods, continuous integration and continuous delivery (CI/CD) of constantly increasing and evolving capabilities, automation in testing and verification, and secure practices throughout the development life cycle—has transformed commercial software development approaches and are increasingly being adopted across the national security community. Similar concepts are needed for rapid and continuous modernization of the range enterprise, to assure connectivity and security, allow integration of capabilities from different ranges into a composable "range of ranges," make productive use of the exponentially increasing amounts of data generated and collected in testing, and effectively integrate M&S with physical testing for all-domain kill chain T&E.

However, the range enterprise and T&E functions lack a comprehensive, flexible, and scalable data strategy, resulting in the inefficient use of data collected currently, and the failure to collect some of the most important data that can be used to inform and optimize operational testing. Ranges often lack the resources, infrastructure, tools, and processes to handle the scope and scale of required data and computational operations, and are typically not operating in a seamless end-to-end digital thread from requirement definition through verification and validation (V&V). Data communication and integration challenges, including limited bandwidth for collection and timely transmission, and incompatible standards and formats for sharing and combining data across ranges and test systems, result in extensive manual efforts and delays in analysis when performing tests across multiple ranges or operational domains. Challenges often beyond the control of the ranges further burden the T&E system, particularly slow, laborious, and manual security approval processes coupled with an absence of distributed, multi-level security information systems.

DoD and the range enterprise must develop a data strategy that emphasizes speed and interoperability, and must further define, adopt and promulgate modern interoperable approaches for seamless, fast and secure collection, transmission, sharing, and analysis of very large data sets. Concurrently, the Director of Operational Test and Evaluation (DOT&E) and the ranges need to incorporate modern software development approaches, enabling a "TestDevOps" construct (for Test-Development-Operations, based on DevSecOps concepts) that leverages digital engineering, permits rapid capability upgrades for software-intensive test systems, and enables effective testing of future systems that will incorporate new software- and data-driven technologies like artificial intelligence and machine learning (AI/ML).

Reinvent the Range Enterprise Funding Model for Responsiveness, Effectiveness, and Flexibility

Given the pace of technological change in U.S. weapon systems, and the relentless and rapid advances in adversary systems across all domains of conflict, speed in testing and learning is critically important for ranges as they seek to modernize their capabilities while sustaining test assets that are often decades old, single-string, and unreliable. A revitalized approach to early definition of range capability requirements is required at both the program and enterprise levels, but that change alone will not address the resource challenges imposed on the range enterprise by today's complex, uncertain, and inflexible approach to funding for operations, sustainment, and modernization.

Today, range capabilities are funded by a variety of streams from individual programs, the military services and DoD agencies, military construction (MILCON), and central pools such as TRMC resources. Funding levels are subject to annual appropriations and vulnerable to out-prioritization by other emerging DoD needs, resulting in great unpredictability from year to year. Much of the funding is limited to specific purposes and cannot be reallocated based on greatest need, while funding often arrives late in the fiscal year and may have unrealistic requirements for timing of obligations and expenditures for test systems that require years of development and construction. Downward trends in MILCON for T&E collide with increasing demands for test time and modernized capabilities.

Based on this funding landscape, some immediate steps are needed to improve the flexibility in timing of spending for range infrastructure, and also increase the ceilings for flexible use of research, development, test, and evaluation (RDT&E) or operations and maintenance (O&M) funding for minor military construction to enhance the valuable authorities already provided by the Congress. More sustained impact will be enabled through the pilot of additional changes to DoD's range funding mechanisms, including establishment of a Range Working Capital Fund to stabilize funding for modernization and sustainment, and demonstration of these changes to prioritize and correct capabilities gaps needed for timely T&E of new systems with multi-domain test requirements.

STRUCTURE OF THE REPORT

The remaining chapters of this report are structured into three broad pieces: an examination of what the future of warfare could look like and the implications of that envisioned future for operational testing and evaluation; three chapters each devoted to the one of the three broad themes of this report (kill chains, digital technologies, and speed-to-field); and a chapter summarizing the report's findings and recommendations. The structure is designed to enable the reader to understand the environmental forces, the state of DoD testing, and the projected future requirements that resulted in the key themes and lines of effort summarized above. The individual chapters provide detail and supporting examples on the themes, the lines of effort, and the recommendations they encompass.

More specifically, Chapter 2, on the envisioned future of warfighting and OT&E, addresses several questions: What is the future of warfare, and what will be required for OT&E? How will testing need to change? What are the implications for DoD's test and training range enterprise? Finally, what does "good" look like?

Chapter 3, on kill chains, multi-domain operations, and the associated future testing needs, begins with an examination of how kill chains and multi-domain operations work. Next it discusses the challenges to testing of systems in operationally relevant ways that arise in tests involving kill chains or carried out over multiple domains, or both. Finally, the chapter examines what the ranges require for carrying out a kill chain approach to operational testing (OT).

Chapter 4, on digital technologies—mainly focusing on modeling and simulation and on the ranges' digital infrastructure—discusses two broad categories of challenges. The first is testing issues related to the increasing role of digital engineering and modeling and simulation in the development of weapon systems; the second is challenges related to the ranges' digital infrastructure, such as the sharing of data among ranges and the corresponding requirements for increased connectivity, interoperability, and security. Among the questions addressed by the chapter are: What does digital engineering mean for test, and why does M&S matter for OT? What needs to change in the range enterprise to take full advantage of digital engineering and M&S? What is needed in the range enterprise for effective OT of software-centric systems? Finally, what range digital infrastructure is required for connectivity and effectiveness?

Chapter 5, on speed-to-field, begins with a discussion of why speed is so critical for OT in today's world. It examines the range enterprise factors that contribute to delays in fielding systems and kill chains and asks what needs to change in the way that OT and range requirements are established. Finally, it surveys the resource challenges that affect OT speed and efficiency and recommended remedies.

Chapter 6 wraps up the main portion of the report by bringing together all of the committee's key findings and recommendations in one place, with the recommendations by stakeholders.

The appendices provide additional detail on the background, context, study approach, and sources of information that informed the committee's work and report outcomes. Appendix A includes the study statement of task. Appendix B provides a summary of the site visits the committee conducted. Appendix C provides biographical sketches of the committee members. Appendix D is the disclosure of unavoidable conflicts of interest. Appendix E lists the abbreviations and acronyms used in this report.

REFERENCES

Andreesen, M. 2011. "Why Software Is Eating the World." *Wall Street Journal*, August 20. https://www.wsj.com/articles/SB10001424053111903480904576512250915629460.

Brose, C. 2020. *The Kill Chain: Defending America in the Future of High-Tech Warfare*. New York: Hachette Books.

Brown, C.Q. 2020. "Accelerate Change or Lose." US Air Force Chief of Staff. https://www.af.mil/Portals/1/documents/csaf/CSAF_22/CSAF_22_Strategic_Approach_Accelerate_Change_or_Lose_31_Aug_2020.pdf.

Cronk, T.M. 2020. "Speed Must Be Put Back Into DoD, Hyten Says." *U.S. Department of Defense–News*, January 17. https://www.defense.gov/Explore/News/Article/Article/2060538/speed-must-be-put-back-into-dod-hyten-says.

Defense Innovation Board. 2019. *Software is Never Done: Refactoring the Acquisition Code for Competitive Advantage*. https://media.defense.gov/2019/Apr/30/2002124828/-1/-1/0/SOFTWAREISNEVERDONE_REFACTORINGTHEACQUISITIONCODEFORCOMPETITIVEADVANTAGE_FINAL.SWAP.REPORT.PDF.

DoD (Department of Defense). 2018. *Summary of the 2018 National Defense Strategy of the United States of America: Sharpening the American Military's Competitive Edge*. https://dod.defense.gov/Portals/1/Documents/pubs/2018-National-Defense-Strategy-Summary.pdf.

DoD. 2020. *Mission Engineering Guide*. Washington, DC. https://ac.cto.mil/wp-content/uploads/2020/12/MEG-v40_20201130_shm.pdf.

GAO (Government Accountability Office). 2021. *F-35 Joint Strike Fighter: DoD Needs to Update Modernization Schedule and Improve Data on Software Development*. Washington, DC. https://www.gao.gov/assets/gao-21-226.pdf.

Mowthorpe, M. 2005. "The revolution in military affairs (RMA): The United States, Russian and Chinese Views." *Journal of Social, Political, and Economic Studies* 30(2, Summer): 137–153.

Murray, W. 1997. "Thinking About Revolutions in Military Affairs." *Joint Forces Quarterly*. https://apps.dtic.mil/sti/pdfs/ADA354177.pdf.

NASEM (National Academies of Sciences, Engineering, and Medicine). 2021. *Key Challenges for Effective Testing and Evaluation Across Department of Defense Ranges: Proceedings of a Workshop—In Brief*. Washington, DC: The National Academies Press.

Wolfe, F. 2021. "F-35 Joint Program Office Expects to Receive Analysis on Feasibility of Joint Simulation Environment." *Defense Daily*, April 22. https://www.defensedaily.com/f-35-joint-program-office-expects-receive-analysis-feasibility-joint-simulation-environment/air-force.

2

An Envisioned Future of Operational Test and Evaluation

In its statement of task the committee was asked to assess the physical and technical suitability of the U.S. military's system of operational test and evaluation (OT&E) and to provide recommendations on addressing any deficiencies in that system relative both to existing technologies and to any technologies expected to arrive over the next decade and a half. To guide itself in that task, the committee developed a vision of what an ideal OT&E system would look like in 2035. Working with this envisioned future allowed the committee to be more methodical and consistent in identifying the ways in which today's military ranges could be improved and helped in developing recommendations for how best to achieve that desired state. While the vision is by necessity incomplete and will certainly have failed to anticipate some future developments, the committee believes that moving in the direction of this envisioned future will produce a future OT&E that is characterized by agility, flexibility, and speed.

This chapter paints a vision for the future that provides foundation and context for the remainder of the report. In the following chapters the committee analyzes in detail the challenges facing today's military ranges, offers its conclusions about the current state of OT&E, and provides recommendations for improving the ranges and preparing them for the challenges they will face in the coming years. Before that, however, it is important to pull back and see the "big picture" of what the nation's military test ranges could ideally become.

The description of OT&E's envisioned future takes place in three steps. The first is a vision of what warfare in the future is likely to look like, particularly warfare with a near-peer or peer adversary. This is an exercise that has been carried out at multiple times by multiple groups, and the committee has relied on such outside work for its vision of the future of warfare. From there, the next step is to envision what a system of OT&E would need to look like to help produce weapon systems that would operate successfully in such conflict. A crucial principle here is "test as you fight"—that is, the testing of weapon systems should take place in an environment that is as close as possible to the environment in which they will actually be used. This principle, combined with a vision of how future warfare will be conducted, leads to a vision of what military ranges should look like in this future. Finally, working from the joint visions of future warfare and future OT&E, this chapter describes an envisioned future for the organizational and funding structures necessary to enable the future ranges and testing. Thus this chapter will lay out an envisioned future in three steps: future warfighting, future military ranges and OT&E, and the future enablers of such testing.

THE FUTURE OF WARFIGHTING

How will wars be fought in the coming decades? Much has been written on this topic from defense analysts and think tank scholars working to anticipate the military's needs. This section focuses on the parts of that future that will pose the greatest challenges to the nation's military ranges.

Novel Weapons and Domains

Weapons and military technologies have steadily become more sophisticated and more effective, but few, if any, eras in history have seen the pace of changes in military systems being experienced today. Not only new weapons, but entirely new classes of weapons are in use or under development, leading to fundamental changes in the nature of warfighting.

Consider hypersonic weapons, which are capable of traveling vast distances through the atmosphere at high Mach numbers and able to maneuver in order to avoid standard anti-missile defenses. Hypersonic weapons have capabilities that are fundamentally different from previously existing weapons and thus require new systems and strategies to counter them (Stone, 2020; Vergun, 2020).

An even more revolutionary transformation is promised by artificial intelligence (AI) and autonomous systems (Hoadley and Sayler, 2020; NSCAI, 2021). Powered by the dramatically increasing speed and power of digital technologies, AI is already applied in a wide variety of areas, both military and civilian, and the number of applications is expected to grow in coming years. In particular, many weapons and military systems in the future will be "smart"—able to carry out many functions without human input. Military applications will range from the rapid processing of intelligence data to the control of autonomous vehicles and systems, which will in some cases be given the independence to make battlefield decisions without direct human control.

Both AI and hypersonics reflect the increasing speed of warfare, and their expected integration may prove very powerful. As stated in a 2019 RAND report, "The pace of war can exceed the speed at which humans can observe what is happening, conceptualize a strategy, and deliver commands" (Winkler et al., 2019, p.13), which will drive the need to develop and test approaches for automated decision making. Although it is impossible at this point to predict exactly which AI technologies will play a significant military role, AI and autonomous systems are already spearheading fundamental shifts in military conflict.

Furthermore, warfighting in the future will be marked not only by new types of weapons, but by new domains of warfare. In addition to the traditional military domains of air, sea, and land, the emergence of the cyber domain as a contested space has been recognized for more than a decade (CSIS, 2021). There have been opening gambits made, such as the covert introduction of programs into the U.S. electric grid, which may have been made in anticipation of launching serious, large-scale cyberattacks in the event of a major conflict (Gorman, 2009). Recent ransomware attacks on the Colonial Pipeline Co. and JBS shut down fuel pipelines and meat packing plants, respectively, highlighting how cyber vulnerabilities create national security threats that extend to the nation's energy and food supplies (Lane, 2021; Williams, 2021). Conflicts in the future are likely to include cyber strikes not only on military forces but on their civilian infrastructure, including communications, power, and transportation. It is even possible that an adversary might engage in such a widespread cyberattack on infrastructure without attacking with more traditional military weapons in the hope of destabilizing an opponent while reducing the risk of triggering a full-scale war (CLAWS, 2020).

Another increasingly contested domain is space. Communication and observational satellites provide precision navigation and timing, which play major roles in modern conflicts, and thus they are targets for adversaries seeking to disrupt an opponent's communications or limit the opponent's ability to monitor multiple locations on or above the earth's surface. This in turn has led to a growing focus on both offensive and defensive weapons that can be deployed in space, with the United States officially creating the Space Force, its first new military service since the creation of the Air Force in 1947, and other countries also placing a new emphasis on space as a domain of military operations (Spirtas et al., 2020). However, as noted at the January 2021 workshop by the acting director of OT&E, Raymond O'Toole, a priority gap for the U.S. Department of Defense (DoD) testing community is the lack of a dedicated range for testing space weapons.

In short, warfighting of the future will involve not just newer and more sophisticated weapons, but new classes of weapons and new domains, signaling the emergence of entirely new approaches to conflict. Furthermore, it is clear that the changes in weapons and military systems in the future will take place at an increasingly rapid pace, bringing new technologies into play at a rate that will be unlike anything that has been seen before. This will be particularly true for digital capabilities, such as AI-enabled systems and the software used in data analysis, command and control systems, other software-intensive aspects of the military enterprise, and even the design and development of new warfighting technologies themselves.

Multi-Domain Operations and Kill Chains

While new weapons and new domains will shape the future of warfighting, an even more transformative factor is the ongoing combination of systems across multiple domains to create an integrated fighting force that is greater than the sum of its parts. The vision for such multi-domain operations is that they will take advantage of rapidly improving communication capabilities and increasing computing power to tie together many different systems, from detection to strike, in such a way that the different pieces act in a coordinated manner, react quickly to changing conditions, and overpower adversaries through the combination of their forces.

The next chapter will dive into deeper detail on multi-domain operations (MDOs), but broadly, the operational view of MDOs is that platforms in different domains share information to accomplish an objective or set of objectives in a given combat scenario. In particular, the kill chain will be apportioned across different elements in different domains, which is in sharp contrast to the traditional kill chain that existed prior to the information revolution. A traditional kill chain is generally contained within a single platform, such as a fighter pilot detecting where an enemy aircraft is, deciding what should be done about it, and then carrying out that action (Clark et al., 2019). In recent years, this traditional kill chain has expanded to include different components; it has become somewhat common, for instance, for data from an observation drone to be sent back to a command-and-control center, which then orders a strike by a stealth fighter. But in the coming years, with the addition of new sensing capabilities, new weapon systems, and new domains of warfighting, the kill chain is only going to become increasingly more complex and sophisticated.

Ultimately, the effectiveness of kill chains and the success of MDOs will depend on how well integrated the various components of an operation are. As stated in the 2018 *National Defense Strategy*, "Success no longer goes to the country that develops a new technology first, but rather to the one that better integrates it and adapts its way of fighting" (DoD, 2018, p. 10).

In short, warfare in the next 15 years or so can be expected to have the following attributes that will make it different from warfare today: Weapons will be more sophisticated, more complex, and more effective. A variety of new technologies will play a role, from hypersonic weapons to autonomous systems. The domains of cyber and space warfare will be part of the picture. The pace of weapons development will continue rapidly increasing, with new technologies brought online more quickly than in the past. And different technologies and domains will be more tightly integrated than in the past.

In the committee's envisioned future, the testing-and-evaluation community will engage in strategic planning efforts along these mission threads, assessing the ability to test both evolutionary and revolutionary advances in technology at a speed that assures continued military advantage. To understand how this works, consider the test and evaluation assessment framework shown in Figure 2.1.

The horizontal axis in this framework indicates testing scale. At the far left is component-level testing that addresses specific subsystems within a given system, or foundational military technology such as radar signal processing. Further along the axis are platforms, such as avionics systems, weapons platforms, and autonomous vehicles. On the far right end of the scale are systems of increasing complexity and number of elements, including the sorts of systems expected in the future, such as human–machine teams operating in consort to achieve mission objectives (Winkler et al., 2019). Current test range capability is well designed to address the component level, and platform level test requirements as new sensors, weapons and vehicle upgrades are worked through the acquisition pipeline (Dahmann et al., 2010). Future test ranges must include this current capability as well as the ability to test larger-scale system of systems.

The vertical axis corresponds to the nature of the new technology under consideration, from incremental improvements to existing capability to novel game-changing technologies that could disrupt military operations. Many of the technologies predicted to come online in coming decades fall into the revolutionary technology category and have the

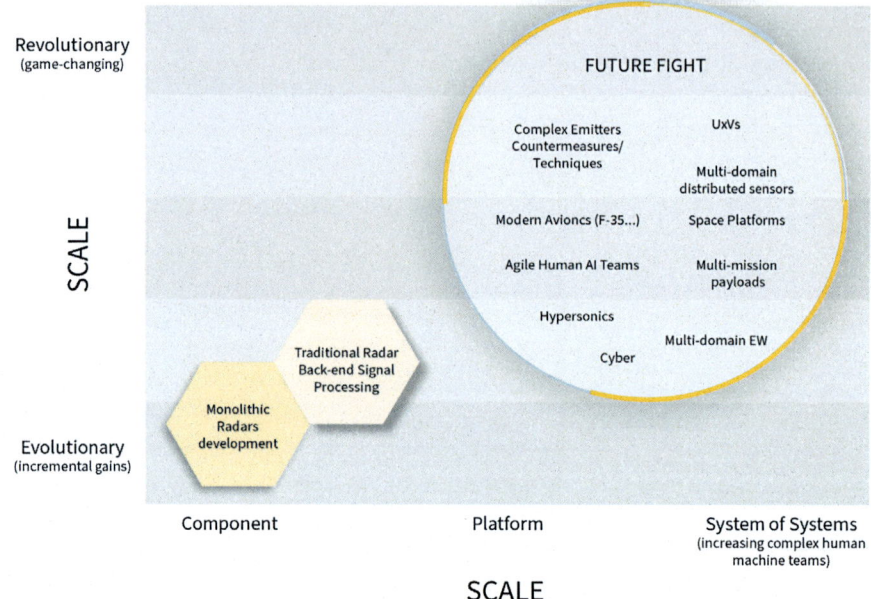

FIGURE 2.1 Proposed test and evaluation assessment framework.

potential to upend military conventions. These capabilities include multi-domain distributed sensors, complex emitters, and hypersonic weapons as well as future cyber- and space-based defensive and offensive systems.

Ensuring that the U.S. military will be successful in future conflicts requires the ability to operate—that is, to effectively carry out testing and evaluation—in the upper right hand corner in this landscape. In the committee's envisioned future, DoD testing and evaluation is a driver in this strategic discussion, working with research and development organizations to explore how a new technology will be tested and carrying that perspective through the system life cycle to streamline the development, evaluation, and fielding of new capabilities.

THE ENVISIONED FUTURE OF MILITARY TEST RANGES

Given this vision of future warfighting, what changes will be required for the nation's military test ranges in order to prepare for it? From its discussions with Test Resource Management Center (TRMC) leadership, DoD stakeholders, leading scientists, and military personnel, the committee contends that a paradigm shift in testing approach will be necessary in order to reach the appropriate future. In particular, ensuring the nation's future warfighting abilities will require integrated capabilities, MDOs, and the seamless adoption of new technologies at the speed of innovation.

The nation's test ranges have never been static, and in coming years their capabilities and infrastructure will need to be refreshed periodically—just as always has been the case. However, the ranges are also facing new challenges unlike any they have faced before, and these will require responses that are also novel. One such challenge is to modernize test range capabilities to match the rapidly increasing pace of technological innovation, particularly in digital technologies such as AI, autonomous systems, and cyber warfare. There are essentially no areas of technology that have been untouched by this accelerating innovation. If the speed of testing and evaluation does not match the speed of innovation, then the testing can serve as a chokepoint, significantly slowing the pace at which new systems can be brought into service and potentially putting the nation at a disadvantage to others who are able to innovate and field faster. Alternatively, testing needs may get neglected, thus passing risk to the operational users and leaving decision makers uninformed about the capabilities and limitations of their systems.

A second—and perhaps more fundamental—challenge will be to test new weapons and systems as part of larger and more complex operations in a way that mimics how the weapon systems will be used in real-world situations. This is not something that has been an emphasis for military ranges in the past. Historically, most of OT&E has been focused on individual weapons and systems and making sure that they work as they are supposed to under conditions similar to those that would be encountered in combat. Although operational plans may involve the use of multiple platforms working together, the capability and understanding of that system of systems integration has often not been a major design or test requirement. With the information revolution, however, the effectiveness of the connections between systems has become more critical than the capability of any one system acting alone.

Given these trends and forces, the committee offers a vision of the military test range of the future (Figure 2.2), and the following discussion delves into some of the details of that vision.

Testing New Technologies

Ideally, by 2035 a close operational testing (OT) and developmental testing (DT) partnership will drive a holistic approach that allows TRMC to address the increasing pace of technological development by developing a set of approaches to testing at a pace that matches that development. For example, TRMC will have collaborated with various research and

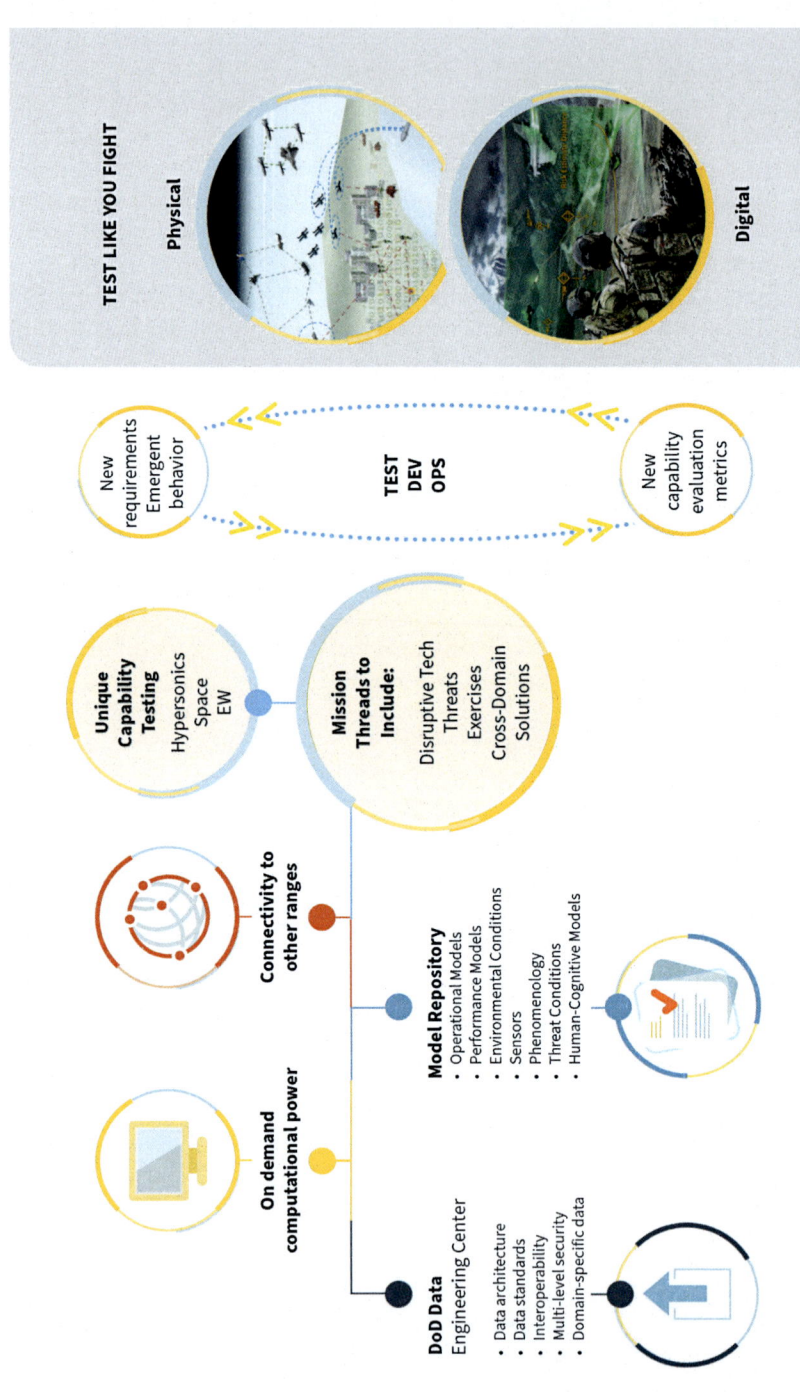

FIGURE 2.2 A notional concept of the military test range of the future.

development organizations within the Department of Defense (DoD)—the Defense Advanced Research Projects Agency (DARPA), the Air Force Research Laboratory (AFRL), the Office of Naval Research (ONR), the Army Research Laboratory (ARL), etc.—to identify effective test approaches and standard metrics that can be carried through the life cycle for developing and evaluating critical emerging and new technologies. This approach may require giving TRMC the authority to hire staff directly, akin to DARPA and other DoD laboratories in order to support the full life cycle view of test planning efforts.

Test ranges will have used a "TestDevOps" approach that connects developmental model-based testing with OT&E real-world scenarios in order to rapidly field and iterate on advances in new technology. A TestDevOps model is one that merges the testing and operational communities in evaluating, refining, and deploying agile solutions to the field. More specifically, TestDevOps takes advantage of the traditional aspects of the DoD test enterprise (i.e., traditional component-level testing to determine if a unit under test has met specific performance requirements plus system-level testing to assess overall use as a DoD capability) into operational scenarios based on advanced modeling and simulation (M&S) as well as real-world multi-domain exercises to determine how a capability will be used as part of a human–machine teamed solution. The user input on the performance of the system in real-world scenarios is then used to inform any system updates and to facilitate release to the field.

In the time between the release of this report and 2035, TRMC and the services will have defined the range requirements for various new capabilities, such as the space systems test range and ranges for testing hypersonic missile technologies, and they will have used that framework as a model to retrofit the test ranges of 2021 to bring them in line with future needs; the requirements in that new framework will have addressed test approaches, computational infrastructure, TestDevOps approaches, the balanced use of M&S versus real-world testing, and networking and data interoperability.

Testing Kill Chains and Multi-Domain Operations

In 2035, as envisioned by the committee, the nation's military ranges will be fully capable of testing kill chains and multi-domain operations, as OT&E is dedicated to the principle of "test as we fight." By 2021 there had already been initial steps toward this end, such as the combined Orange Flag–Black Flag large force test event that took place on March 2–4, which allowed "for improved integration and the combining of resources and participants to provide better test data and a more robust operationally relevant environment" (Saunders, 2021). An Emerald Flag event was introduced in December 2020 which provided a realistic operating environment linking ground, air, and space systems together to demonstrate joint and multi-domain operational capabilities while identifying tangible shortcomings to these systems (Rodriguez, 2020). A look at a few of the details of that operation offers an indication of what is involved in that operation.

Orange Flag test events are focused mainly on technical innovation and integration, and that particular event in March 2021 examined integrated kill chains—or "kill webs," as they were referred to—which involved sensors and tactical networks from the Air Force, Army, Navy, Marine Corps, and Space Force connected via current command and control capabilities that will evolve into future Joint All-Domain Command and Control (Saunders, 2021). Black Flag events, by contrast, focus on ways to improve tactics—"tactic improvement protocols"—for existing weapons and systems. In the joint test event, Orange Flag and Black Flag "combined their mission planning processes and streamlined test objective synthesis," although the execution of the Orange Flag and Black Flag tests was actually separate (Saunders, 2021). The Emerald Flag events are multi-domain test exercises that incorporate technology and prioritize efficiency for joint warfare through rapid data-driven analysis.

Over the next decade and a half, according to the committee's vision, this sort of testing of end-to-end kill chains over multiple domains will have become much more commonplace and much more highly integrated. "Testing as you fight" will require the testing of multiple integrated technologies acting in concert in a way that closely mimics what would happen in real combat.

It should also be noted that, in this envisioned future, the data from these tests of kill chains and multi-domain operations will be collected and analyzed in such a way that makes it possible to pinpoint weaknesses and failures—which can actually be quite difficult in integrated tests involving multiple components interacting in complex ways. This approach will prove valuable for high level operational test objectives, and will require further iteration when complex technical calculations and required to configure test instrumentation. Furthermore, because of the value of M&S to these sorts of tests (see next section), OT&E will have implemented a policy that a key objective of every end-to-end kill chain test is to produce data for use in the validation of simulation models and dynamics.

Modeling and Simulation

In the envisioned future, as today, it will not always be practical to carry out physical tests of a technology. In some cases—such as a nuclear bomb or a weapon designed to disable computers or communications—it is simply too dangerous or disruptive to realistically test military technologies. In other cases it may be critical to control what adversaries observe about U.S. military systems in action. In still other cases it may simply be too expensive to carry out the number of tests required for a full evaluation of a technology. In all of these cases and others, simulated tests using a model of the technology is often the best option. And, with the increased computing power available in the future and the more complex and sophisticated technologies that must be tested, M&S will play a much more integral role in OT&E.

By 2035, according to the committee's vision, M&S will have been integrated into nearly all operational test and evaluation activities. While the promise of M&S capabilities has fallen short in some attempts to date, the science continues to evolve and improve and it is necessary that M&S remains a DoD focus area committed to making this approach work. There will be a tight interplay between physical testing and M&S, with the physical tests providing data to guide the development of or validate the models, and the M&S indicating which particular aspects of a system should be tested and what data should be collected. Planning for M&S will begin early in the concept-development phase in order to support design decisions during development and subsystem integration, with later integration into developmental and operational test programs, campaign-level exercises, and, eventually, operational sustainment. The models will have evolved over the course of the program and will be adaptable for use at different levels within DoD, providing the required capability at each. Early digital engineering development in the form of model-based systems engineering (MBSE) will address a system's overall requirements, structure, behavior, and data input and output interactions that will aid in viewing composability of a system—a system design principle that deals with the interrelationships of components (SBIR, 2014). A highly composable system provides components that are selected and assembled in various combinations to satisfy specific user requirements. Models within simulations will be developed that are consistent with the MBSE development, resulting in M&S that is relevant throughout the T&E life cycle.

Furthermore, M&S will not be used just in the testing of individual technologies, but also for integrated systems. Models will be used to emulate other technologies that interact with the system under development or elements of the system under development that will not be fully functional in time to exercise other subsystems on the critical path. Certain models that are used over and over again, such as models that represent DoD infrastructure or adversary equipment or threats against which multiple development systems will be tested, will be maintained as part of a common test infrastructure. More generally, there will be a widely shared and accessible M&S ecosystem that includes common scenarios, models, and data that can be used by concept developers, requirements developers, research and development (R&D) programs, and acquisition programs. This M&S ecosystem will have spread across not just DoD but also industry, so that industry models can be used by the military and DoD models can be used to support digital engineering by industry partners.

Also, in this envisioned future, M&S will be carried out according to a set of widely accepted standards that address data interoperability and other issues that result from the growing complexity of the systems being modeled. The M&S architecture can adapt to the evolution of testing programs, including uncertainty quantification. The future must include the ability to design, develop, and test data storage at unprecedented speeds in trusted cross-domain architectures at TTS/SCI/SAP (Top Secret/Sensitive Compartmented Information/Special Access Program) down to unclassified levels. Chapter 4 provides a deeper rationale behind this vision.

Data Sharing, Repositories, and Accessibility

In the envisioned future, military ranges and other development and testing facilities will have the bandwidth and connectivity to share data and models and enable rapid data analysis across multiple classification levels. To effectively handle data from testing, program managers work early in the design phase with the ranges to develop a data strategy to inform operational testing. Some of the improvement will be due to the presence of a properly curated and protected data and model repository that will be widely accessed by those in the testing and evaluation community. A second factor is the development of data protocols for the real-time transfer of data at appropriate classification levels and the increased interconnectivity of ranges.

More generally, the T&E system will have the instrumentation, telemetry, data collection, data handling, and data analysis capabilities necessary to collect, transfer, store, and manipulate the huge amounts of data that are generated by the many different types of sensors observing tests. The capacity will exist to handle new and emerging types of data-collection systems, including those that are not stationary, such as instruments about wave runners, aircraft, and satellites. Systems will have been developed to collect and transfer data from tests carried out in integrated environments across multiple ranges.

ENABLING THE ENVISIONED FUTURE OF MILITARY RANGES

Given this vision of the future of OT&E—which is significantly different from the present version of OT&E—what sorts of changes will have been made by 2035 in policies, requirements, acquisitions, and funding to enable this new approach to testing? The committee envisions many such changes.

Critical Joint Mission Threads

In the envisioned future of 2035, it will be understood there are a few mission threads so important to DoD that their execution should be tested in a coordinated fashion. These are referred to as "critical joint mission threads," which are end-to-end sets of activities and systems that accomplish the execution of a joint mission.[1] Most future advanced technologies are integral to these critical joint mission threads and therefore require a new perspective on testing and evaluation, including how tests are planned, funded, and used to drive new technology into the field.

The test planning for these critical joint mission threads will not be owned by a program, but by a larger organization empowered by the J-8 Directorate of the Chairman of the Joint Chiefs of Staff with the responsibility for integrated test and evaluation for that particular mission thread. This "joint program office" will have a larger set of responsibilities beyond integrated testing: It will serve as the certifying authority for DoD's capability to perform critical joint mission threads, it will perform mission engineering for its mission thread, it will develop an authoritative operational and top-level systems architecture for the mission, it will develop and analyze integrated requirements for the participating systems, and it will inform acquisition decision milestones for participating programs.

Essentially, such a joint program office will use testing and evaluation as a mechanism to foster agile acquisition and development in its mission area. To support agile "TestDevOps," the joint program office will drive an integrated set of development and test activities, integrating live test and modeling and simulation to take advantage of the strengths of each. It will steward and fund persistent range capabilities and M&S repositories, along with a test data repository, and it will enable access to these repositories across the services, programs, and industry partners.

Funding and Acquisition

The envisioned future for T&E will have funding and acquisition mechanisms that represent improvements in various ways over the current system, such as a resolution to complex and disconnected funding

---

[1] "Critical joint mission threads," as defined in the Defense Acquisition University glossary, https://www.dau.edu/glossary/Pages/Glossary.aspx#!both|J|27776, accessed June 16, 2021.

streams; flexibility to apply investments from different sources to priority needs; predictable multi-year funding with adequate authorities to obligate/expend at the right timing for T&E needs; adequate funding for ongoing sustainment and modernization of existing capabilities; and funding approaches that are well suited to software-intensive test systems. Chapter 5 recommends a pilot program to build closer cooperation between the OT and DT communities that includes a new process for funding ranges and infrastructure to make it simpler, more responsive, and more effective. Still, there are a few things that can be confidently stated about the envisioned future of funding and acquisition.

For example, as part of the coordination between development and testing, DOT&E will be included early in the acquisition process in order to coordinate testing requirements and to collaboratively identify shortcomings in testing capabilities. Furthermore, T&E funding streams will be established early in the development process to ensure the ranges will be ready to do appropriate testing when the system being developed is ready to be tested, and operational and developmental testing requirements will be synchronized early in the acquisition process. Additionally, TRMC will have set out a mutual capability requirement process that is more responsive to emerging technology testing than it was in 2021.

The process for funding testing and evaluation ranges and infrastructure will be simpler in this idealized future, and barriers to range modernization will have been identified and, where possible, removed. Better mechanisms will have been developed for funding range maintenance and the development of new capabilities to meet emerging technology requirements. There will be a working capital fund for the ranges. And better funding mechanisms will have been identified for software-enabled capabilities and the maintenance of software over time.

In the specific area of M&S, in the envisioned future, requirements for the full hierarchy of M&S to support a system through its entire life cycle will be accounted for and funded during early concept development. M&S will be persistent so that it supports nearly all life cycle activities, from concept and requirements development through operational testing and sustainment. To enable this, the necessary resources will have been provided to support M&S with a stable funding profile, and requirements will be established for specific M&S capabilities to support development decisions and integration with the test program.

Finally, given the importance of testing kill chains and multi-domain operations and the lack of any natural home for these activities, the envisioned future includes a joint program office to support connected concurrent kill chain operations as an OT&E activity. This activity provides support for enterprise-level simulation environments; identifies dedicated funding for the development, sustainment, and management of T&E data and a model repository; and includes test infrastructure for assessing the integration of a new capability as part of the development of that new capability, not as a separate effort. Such a system ensures that range capabilities can support the operational assessments of concurrent kill chain operations.

Mitigating Range Encroachment

In the envisioned military range system of the future, range encroachment will continue to be a concern and constant issue, but there will be improvement in the use of mitigations that will preserve DoD test capability. DoD will preserve dedicated frequencies for weapon and threat testing, and will have more capability to efficiently manage and utilize dedicated spectrum and have capability to continue some test operations in areas of shared spectrum. The potential for physical encroachment will still exist but issues of internal encroachment will be addressed and managed to prevent impact to test operations. And better range management will have led to fewer test failures caused by issues with emitters from other tests.

This envisioned future offers a preview of the remainder of the report. In the following chapters the committee offers its analysis of the current state of the systems of military ranges, including both its strengths and weaknesses, describes specific changes that could be made to improve those ranges and prepare them for 2035, and then provides specific recommendations on how those changes could be accomplished. But the end goal of those recommendations is right here in this chapter—with the vision of what OT&E could be and should be a decade and a half from now.

REFERENCES

Clark, B., D. Patt, and H. Schramm. 2019. "Decision Maneuver: The Next Revolution in Military Affairs." *Over the Horizon Journal*, April 29. https://othjournal.com/2019/04/29/decision-maneuver-the-next-revolution-in-military-affairs/.

CLAWS (Center for Land Warfare Studies). 2020. "The Shades of Cyberwarfare in the Era of Grey Zone Conflicts." *CLAWS Focus*. https://www.claws.in/the-shades-of-cyberwarfare-in-the-era-of-grey-zone-conflicts/.

CSIS (Center for Strategic and International Studies). 2021. *Significant Cyber Incidents*. https://www.csis.org/programs/strategic-technologies-program/significant-cyber-incidents.

Dahmann, J., J.A. Lane, G. Rebovich, and R. Lowry. 2010. "Systems of Systems Test and Evaluation Challenges." Pp. 1-6 in *5th International Conference on System of Systems Engineering*. doi: 10.1109/SYSOSE.2010.5543979.

DoD (Department of Defense). 2018. *Summary of the 2018 National Defense Strategy: Sharpening the American Military's Competitive Edge*. https://dod.defense.gov/Portals/1/Documents/pubs/2018-National-Defense-Strategy-Summary.pdf.

Gorman, S. 2009. "Electricity Grid in U.S. Penetrated by Spies." *Wall Street Journal*, April 8. https://www.wsj.com/articles/SB123914805204099085.

Hoadley, D.S., and K.M. Sayler. 2020. *Artificial Intelligence and National Security*. Congressional Research Service Report R45178. https://crsreports.congress.gov/product/pdf/R/R45178/10.

Kushner, D. 2013. "The Real Story of Stuxnet." *IEEE Spectrum*. February 23. https://spectrum.ieee.org/telecom/security/the-real-story-of-stuxnet.

Lane, S. 2021. "JBS Attack Unlikely to Cause Major Meat Disruption: USDA." *The Hill*, June 3. https://thehill.com/policy/finance/556696-jbs-attack-unlikely-to-cause-major-meat-disruption-usda.

NASEM (National Academies of Sciences, Engineering, and Medicine). 2018. *Multi-Domain Command and Control: Proceedings of a Workshop—In Brief*. Washington, DC: The National Academies Press.

NSCAI (National Security Commission on Artificial Intelligence). 2021. *Final Report*. https://www.nscai.gov/wp-content/uploads/2021/03/Full-Report-Digital-1.pdf.

Rodriguez, K. 2020. "Emerald Flag Exercise Begins." *Wright-Paterson AFB News*, December 1. https://www.wpafb.af.mil/News/Article-Display/Article/2432103/emerald-flag-exercise-begins/.

Saunders, C. 2021. "Test like We Fight: 'Orange Flag,' 'Black Flag' Collaborate to Accelerate Change." *Edwards News*, March 8. https://www.edwards.af.mil/News/Article/2537319/test-like-we-fight-orange-flag-black-flag-collaborate-to-accelerate-change/.

SBIR (Small Business Innovation Research). 2014. "Predictive Modeling Tools for Metal-Based Additive Manufacturing." Department of Commerce. https://www.sbir.gov/content/predictive-modeling-tools-metal-based-additive-manufacturing-0.

Sick, A. 2018. "Looking Beyond Your Service for Multi-Domain Success." *Over the Horizon*. December 24. https://othjournal.com/2018/12/24/oth-anniversary-looking-beyond-your-service-for-multi-domain-success/.

Spirtas, M., Y. Kim, F. Camm, S.M. Ross, D. Knopman, F.E. Morgan, S.J. Bae, M.S. Bond, J.S. Crown, and E. Simmons. 2020. *A Separate Space: Creating a Military Service for Space*. RAND Report RR-4263-AF. Santa Monica, CA: RAND Corporation. https://www.rand.org/content/dam/rand/pubs/research_reports/RR4200/RR4263/RAND_RR4263.pdf.

Stone, R. 2020. "'National Pride is at Stake': Russia, China, United States Race to Build Hypersonic Weapons." *Science*, January 8. https://www.sciencemag.org/news/2020/01/national-pride-stake-russia-china-united-states-race-build-hypersonic-weapons.

Vergun, D. 2020. "Shortfalls of Defensive Hypersonic Weapons Must be Addressed, NORAD General Says." *DoD News*, October 29. https://www.defense.gov/Explore/News/Article/Article/2399093/shortfalls-of-defensive-hypersonic-weapons-must-be-addressed-norad-general-says/.

Williams, B.D. 2021. "Colonial Pipeline Cyberattack Follows Years of Warnings." *Breaking Defense*. May 10. https://breakingdefense.com/2021/05/pipeline-cyberattack-follows-years-of-warnings/.

Winkler, J.D., T. Marler, M.N. Posard, R.S. Cohen, and M.L. Smith. 2019. *Reflections on the Future of Warfare and Implications for Personnel Policies of the U.S. Department of Defense*. RAND, Washington, DC. https://www.rand.org/pubs/perspectives/PE324.html.

3

Testing for Future Combat: Multi-Domain Operations, Connected Concurrent Kill Chains, and Mitigating Encroachment

Critical to success in a dynamic warfighting environment is the seamless integration of multiple systems and technologies working in concert across multiple domains (land, air, sea, cyberspace, and space); therefore, it is necessary that the operational effectiveness and suitability of emergent technologies are tested in such an environment so that they can be applied to their greatest effect. This critical point was articulated when former Department of Defense (DoD) Director of Operational Testing and Evaluation (OT&E), Hon. Robert Behler, addressed the committee at their opening meeting in December 2020:

> Everybody is going have access to the weapons that we have . . . the trick is, how do we put that together in combined arms? To be able to integrate it all together so that (1) our weapons will be better, and (2) we'll know how to integrate and fight together. How do we put it all together? Our ranges have to be able to compensate.[1]

Historically, OT&E has focused on the performance of individual programs and systems by making sure that they achieve desired outcomes under conditions similar to those that would be encountered in combat. Although operational plans may involve the use of multiple systems working together, their collaborative effects are not typically a major test

---

[1] From remarks delivered at December 4, 2020, committee meeting; recording available at https://www.nationalacademies.org/our-work/assessing-the-physical-and-technical-suitability-of-dod-test-and-evaluation-ranges-and-infrastructure.

requirement, nor are the results of large scale and integrative tests fed back into a program's design. Consequently, if operational issues arise during a multi-system test, there may not be a mechanism to use those results to modify a system's design.

Integration is an increasingly important aspect of military weapons and systems. As stated in the 2018 National Defense Strategy, "Success no longer goes to the country that develops a new technology first, but rather to the one that better integrates it and adapts its way of fighting" (DoD, 2018a, p. 10). The increasing interconnectedness and complexity of systems-of-systems is becoming the operational norm and better aligns with the concept of "testing like we fight" than does testing systems separately.

The requirement that systems be able to work collaboratively to satisfy mission objectives results in a need to represent a variety of systems, both friendly and adversary, under test conditions. A relevant example of this system collaboration was reflected in the 2020 Director of Operational Test and Evaluation (DOT&E) Annual Report, which recognized that in the case of a National Space Test and Training Range (NSTTR), currently referred to as a National Space Test and Training Complex, operationally representative threats must simultaneously include "cyber, directed-energy, kinetic, and electronic-warfare threats, as well as natural hazards." (DOT&E, 2020, p. 3). Additionally, understanding the impact of emerging technologies on mission accomplishment is only understood in the context of its value added to collaborative effects. Measuring and evaluating collaboration between systems as well as the effectiveness and suitability of the resulting system-of-systems (SoS) has become more operationally relevant than testing the capability of any one system in isolation.

The emphasis of this chapter is on the range capabilities needed to assess these systems-of-systems working collaboratively across multiple domains (land, air, sea, space, and cyberspace) and across multiple technologies. The context for this chapter is about improving test range capabilities to support the assessment of multiple concurrent kill chains of systems and technologies and how those systems are connected together within a command-and-control structure to verify and understand the combat effectiveness and suitability within a multi-domain environment. The understanding of kill chains is paramount in assessing a system's operational effectiveness and suitability as well as the integration of new systems into existing kill chains.

TESTING FOR THE MULTI-DOMAIN BATTLESPACE

The multi-domain battlespace can be represented in a variety of operational views. The operational view depicted in Figure 3.1 is a

FIGURE 3.1 The multi-domain battlespace, including connected systems across land, air, sea, space, and cyberspace domains. Red icons denote enemy systems.

representation of how different systems in different domains share information to accomplish an objective or set of objectives.

In the future multi-domain battlespace, "shooters" and sensors both collect data, share that data, information is derived from that data, the data prompts a timely human or autonomous decision, and an appropriate effects-based action results from the data-driven decision. This collaboration of systems represents what the range infrastructure, to include virtual range infrastructure, must support moving forward. DoD ranges must be able to connect with each other, as they gather and analyze data to verify these assumptions, to inform system designs, digital models, acquisition decisions, tactics, techniques and procedures (TTPs), and operational employment decisions.

This committee calls out specific range capabilities and additional enterprise needs that enable these range capabilities in supporting multi-domain and multiple concurrent kill chains. While some of these range capabilities have been highlighted in various reports, the committee paid particular attention to the challenges raised by test range personnel in the site visits and program representatives from the public workshop. The needed test range capabilities to support operational testing of multi-domain systems and multiple concurrent kill chains include:

- High-bandwidth connectivity across ranges, with multi-level security provisions, and common data standards for interoperability (Eglin, Edwards, and Wright-Patterson Air Force bases; Missile Defense Agency [MDA]; Atlantic Test Range at Patuxent River;

Point Mugu; Nevada Test and Training Range [NTTR]) (NASEM, 2021).
- An overarching, cross-range data strategy, processes, and procedures for collecting, storing, managing, and sharing test data (Aberdeen, MDA, Atlantic Test Range at Patuxent River) (NASEM, 2021).
- Capabilities and success criteria for measuring and evaluating collaboration among systems, and end-to-end SoS performance (Eglin, Edwards, and Wright-Patterson Air Force bases).
- The emulation of physical or threat environments that could affect closure of the kill chain in an operational setting (Joint Simulation Environment) (NASEM, 2021).

The enabling enterprise needed to support the above capabilities includes:

- The identification of a process and oversight body for defining kill chain and multi-domain operation (MDO) doctrines and concepts of operation as well as creating cross-program and multi-system test requirements to ultimately drive range capability requirements.
- A defined funding approach to support the execution of "beyond program" multi-domain and multiple concurrent kill chain testing.
- A defined funding approach for sustaining the MDO/kill chain joint infrastructure on the test ranges to last beyond the program funding that originally built the capability.

As DoD advances capabilities in areas such as hypersonics, directed energy, cyber, and artificial intelligence, there will be aspects of multi-domain effects that are essential for understanding how DoD can use these technologies in concert to achieve a desired outcome.

Defining Multi-Domain Operations

Warfighting has long involved multiple domains. For example, Union troops used balloons in the Civil War to help direct artillery (American Battlefield Trust, n.d.). What is different today is how capabilities in different domains are tightly integrated, how much more effective operations can be by taking advantage of such integration, the speed at which information is exchanged, and how new technologies affect the effectiveness of this integrated capability.

In order to modernize range infrastructure to support the operational testing objectives of connected systems in the multi-domain battlespace, it is necessary for the DoD services to agree on how to define MDOs. In this context the committee refers to the term MDO as a more general description of the concept, rather than the Army's vision or the joint vision (joint all-domain operations), both of which have appeared with increasing frequency over the past decade. MDO describes operations that extend over more than a single war-fighting domain—land, sea, air, cyber, and space—although the term has been used in other ways as well (Grest and Heren, 2019). For example, a ground mission with air support or the use of a satellite to guide munitions dropped from an airplane are considered MDOs (NASEM, 2018). MDOs can involve multiple platforms, multiple technologies, and the command-and-control systems that enable that integration across platforms or technologies. A true MDO is a tightly integrated combination of different technologies from different domains under command and control that results in a unified war-fighting operation.

Much that has been written over the past few years about MDOs has been in the context of multi-domain command-and-control, with little said about operations (Grest and Heren, 2019). Thus, many references to MDOs in military publications are actually referring to multi-domain command-and-control systems. In today's military, however, integrated effects from multi-domain systems and command-and-control systems are so tightly coupled that there is little to be gained from distinguishing a difference between them. For example, an F-35 can be considered either a sensor or a shooter, depending on the situation at that moment. At this time, DoD has no formal definition for MDOs but there are multiple interpretations and applications of MDOs.

The Army has developed language to define how MDOs pertain to their domain. It defines MDOs as how they "as part of the joint force [Army, Navy, Air Force, and Marines] can counter and defeat a near-peer adversary capable of contesting the U.S. in all domains [air, land, maritime, space, and cyberspace] in both competition and armed conflict" (CRS, 2021a). Having a defined outcome for MDOs helps clarify strategies to meet Army objectives and provides guidance for congressional oversight. However, the committee noted that the Army's definition does not specify what constitutes MDOs.

Developing a shared DoD vision on MDOs could align the services in a coordinated approach to ensure that corresponding investments are made in systems needed to successfully test in MDOs. Without a clear definition of MDOs, it is challenging to focus the test and evaluation (T&E) investment strategy to modernize range infrastructure in support of MDO testing. Using MDO objectives and the test parameters that accompany a program's life cycle may help broaden the currently program-centric acquisition process as well, shaping program requirements and milestones that better align with mission objectives.

The lack of a common definition for MDO has been frequently cited as a challenge to joint force efforts, which are critical for coordinating the services as they work to deter and win future conflicts (NASEM, 2018; CRS, 2021a). A common definition for MDOs will also provide improved coordination for joint allied efforts, such as joint targeting to synchronize fires with multiple military capabilities across allied nations (NATO, 2016). In an effort to develop a common definition, the committee highlights a definition shared at a National Academies of Sciences workshop in November 2018 on multi-domain command and control by Brig. Gen. B. Chance Saltzman (U.S. Air Force) that suits the complex multi-faceted nature of this term:

> MDOs are more than just assets in one domain participating in operations in another . . . [but] the seamless integration of assets in all domains to create effects in any domain that presents challenges for adversaries that must be addressed in all domains. In effective MDO the need for information or effects in one domain can be achieved through any domain and can complement information and effects from the other domains through seamless integration of platform capabilities and technologies (NASEM, 2018).

Multi-domain operations require integration of capabilities in different domains, effectiveness of operations from this integration, and speed of information exchange. These capabilities are achieved by new systems that closely integrate hardware and software while implementing new technological advances. These new types of systems create a new type of system complexity that requires further definition.

Defining Cyber-Physical Systems

A key comment that the director of DOT&E shared when he addressed the committee was that DoD has no definition of complex systems that have both hardware and software components and that software is a major aspect of almost all new weapon systems. The term "cyber-physical system" (CPS) captures the integration of these technologies. CPSs are complex systems consisting of both hardware and software components. This term deserves particular consideration because of its growing relevance to military systems. The F-35 Joint Strike Fighter and AEGIS systems are good examples; both are software-intensive systems with significant platform integration. Both have real-time command and control communications inherent in the successful performance of their desired effects as well as a reliance on other systems to provide sensor information for adequate situational awareness for successful completion of kill chains.

The complex interactions between software and hardware can sometimes be difficult to predict and pose challenges to operational assessment and range capabilities. For example, maintaining positive control and range safety for emerging non-deterministic, learning artificial intelligence (AI) systems is a challenge for test ranges. Aegis' cooperative engagement capability (CEC) takes advantage of digital communications to enable air, land, and sea forces to share target information in real time; however, software updates to this system can pose challenges to the overall effective operation of the system-of-systems.

While DoD does not currently have a formal definition of CPS, the National Science Foundation issued the following definition in its program solicitation:

> Cyber-physical systems (CPS) are engineered systems that are built from, and depend upon, the seamless integration of computation and physical components. Advances in CPS will enable capability, adaptability, scalability, resiliency, safety, security, and usability that will expand the horizons of these critical systems.[2]

This definition could be further expanded for defense purposes to include small and closed systems, such as an on-board oxygen generation system, or a very large, complex, and interconnected system, such as for a networked system on a multi-domain battlefield. Such a definition emphasizes the predominance of software in military systems today and highlights the fact that any combined test, experiment, or exercise for MDO, at its core, is looking at the data connections and vulnerabilities of those connections in understanding mission capability, as illustrated by the F-35 Joint Strike Fighter (JSF) and Aegis systems examples.

Finding 3-1: DoD has no consistent and clear definition for multi-domain operations or for complex systems that have both hardware and software components.

Conclusion 3-1: The lack of a DoD or joint publication set of definitions for multi-domain operations and cyber-physical systems can result in different operational use cases.

Testing Kill Chains

The "kill chain" is a DoD term describing a process of military engagement. Christian Brose's 2020 book *The Kill Chain* describes it as

---

[2] National Science Foundation, 2021, Solicitation 21-551, Cyber-Physical Systems (CPS), https://www.nsf.gov/pubs/2021/nsf21551/nsf21551.htm.

"gaining understanding about what is happening . . . making a decision about what to do . . . [and] taking action that creates an effect to achieve an objective" (Brose, 2020, p. xviii). There are different models in use to describe the kill chain; one common model is F2T2EA (find, fix, track, target, engage, assess), but foundationally the processes are the same (Tirpak, 2000). Operational tests can examine how a system integrates into a kill chain and how information is received, processed, and used to create the desired effect.

Developmental test objectives for complex systems often drive significant instrumentation requirements with large amounts of data required to understand how a system behaves. Operational test objectives often have the additional challenge of gathering data across multiple systems as part of an operational environment or chain of events. These developmental and operational test objectives inform test infrastructure requirements for the ranges. However, emerging military technologies in areas such as directed energy weapons and hypersonic missiles are increasing the physical and technical demands on the nation's test ranges that affect the ranges' abilities to successfully conduct operational testing that examines a full engagement kill chain.

The committee separated the assessment of test range capabilities from the perspective of testing a kill chain around a single system and then the additional challenge of a full multi-domain test, which involves the convergence of multiple kill chains across several systems. Traditionally, programs test parts of a kill chain in isolation in order to satisfy their programmatic decision needs. For instance, a test may focus on whether a given target could be identified and located, initiating a decision to attack it with a particular weapons system. In this case, the objective of the test is understanding if the weapon functions as intended. This focused testing is important, but to ensure operational effectiveness it is also crucial to test the entire kill chain as an integrated system to look for weaknesses in how the various pieces of the kill chain fit together.

As an example, for an aircraft-mounted high energy laser system a developmental test objective could be to assess the performance of the system from the power output compared to the design requirement against a given target. An operational kill chain test objective would examine how that directed energy system receives target information from different sensors, determines decisions to engage a target, interfaces with an operator, understands the engagement itself, and uses information to assess the effectiveness of that engagement. The operational test would also represent the physical and threat environment the system under test could encounter.

Figure 3.2 illustrates a kill chain testing scenario. In the figure, a friendly aircraft targets an enemy (red) ballistic missile transporter erector

FIGURE 3.2 A representation of a realistic kill chain testing scenario in the multi-domain battlespace. A denotes potential transported erector launcher; B denotes enemy representative radars; C denotes enemy aircraft.

launcher (TEL) through the MDO connected, concurrent kill chain cyber-physical system construct. Examination of this scenario through an observe-orient-decide-act (OODA) kill chain framework, the test evolves as follows:

Observe:
- A reconnaissance drone operated by forward deployed special operations team finds, locates, and transmits the TEL information (labeled A) to the joint command and control (C2) network.
- A reconnaissance drone, potentially operated by forward Special Operations Forces (SOF), or even as part of an automated C2 system orchestrating the find, fix, track, target (F2T2) activities of multiple sensors across domains.

Orient:
- Space and aerial electro-optical and infrared (EO/IR) sensors are tasked and attempt to identify viable time critical TEL target and, once it is identified, to provide information to the C2 network.
- Space sensor assets are tasked and, aided by EO/IR and other sensors, track of target is obtained among representative ground clutter.
- Geo-location is handed off to F-35 and C2 network.
- A land- or sea-based radar emitter provides signal for F-35 passive radar with synchronization of emitter and receiver over the network.

- F-35 tracks red TEL via passive radar among representative clutter.

Decide:
- Advanced C2 framework gathers sensor data, assesses available capabilities, and designated authority makes decision to employ F-35 to take action to strike TEL.

Act:
- F-35 deploys a small swarm or kinetic weapon while wingmen execute non-kinetic electronic or cyber-attack to confuse or degrade radar operations long enough for strike to be successful.
- Cognitive electronic warfare (EW) jamming of enemy representative radars (labeled B) occurs throughout mission, requiring active deconfliction with blue communications and radar signal.

Post/concurrent kill chain and assessment:
- Fighter aircraft (Air Force or Navy) intercept and destroy enemy aircraft (red outlined aerial targets in Figure 3.2, labeled C) over the critical area of a combat zone.

The MDO connected, concurrent kill chain, cyber-physical system construct requires that test planning accounts for more dynamics and blue and red force assets than traditional testing. In addition to the red force assets described in Figure 3.2, the scenario requires space-based sensors, airborne sensors, airborne jamming, land- or sea-based radar emitters, strike aircraft, and space-based communications assets be considered in test range planning.

Some range-based capabilities required for the MDO connected, concurrent kill chain cyber-physical system construct example are that Range 1, Range 2, and the Virtual Range, as depicted in Figure 3.2, are connected with adequate bandwidth, availability of type and quantity representations of red capabilities, blue and red force monitoring for truth data, range coordination command and control, adequate distance for weapon type, and adequate electromagnetic spectrum for communications, radar, and jamming. Some range measurement capabilities are also required for sensor performance, communication performance, command and control performance, weapon effects, environmental factors and situational awareness of blue forces. These combined capabilities are currently limited for operational testing.

There are additional limitations to DoD test range capabilities for conducting end-to-end testing of kill chains. In testimony provided to the committee, Lt. Gen. Neil Thurgood, Director for Hypersonics, Directed Energy, Space and Rapid Acquisition in the Office of the Assistant Secretary of the Army, shared how a lack of a secure communications network among the test ranges is one of several challenges that constrains testing for hypersonic programs. Since hypersonic vehicles can travel thousands of miles, multiple test ranges need to collaborate throughout a vehicle's trajectory. Thurgood pointed out that the ranges were not originally developed for concurrent and collaborative testing; most ranges are not connected via secure communication lines and there is a lack of common processes or procedures for ranges to collect, store, manage, or share test data.

In a virtual site visit to the Eglin/Edwards/Wright-Patterson Air Force bases, representatives agreed that the ranges face rapidly growing operational needs to conduct end-to-end and concurrent kill chain testing. In future combat, systems will need to connect and interact with other systems across multiple domains and in a multi-player environment. This need was one of the factors that led to the creation of the Emerald Flag exercise in 2020. Emerald Flag exercises, which to date have been conducted at Eglin Air Force Base, provide a realistic operating environment linking ground, air, and space systems together to demonstrate joint and multi-domain operational capabilities, while identifying tangible shortcomings to these systems.

Representatives from Eglin/Edwards/Wright-Patterson shared how Emerald Flag is a promising example of how ranges can test and evaluate connected and concurrent kill chain reactions. However, these exercises both produce and require large volumes of test data, and most test range infrastructure is currently inadequate for conducting simultaneous data-intensive activities. A further challenge to testing concurrent kill chains is combining data across multiple levels of security. This is especially the case for providing real-time data, since several ranges reported challenges with data sharing both from a policy perspective on what data is exchanged and technical perspective in terms of sufficient bandwidth. Chapter 4 provides a more detailed description about the data challenges facing the test ranges.

While the Emerald Flag exercises bring testing a step closer to "testing like we fight," these exercises are rare, and test ranges do not currently have the infrastructure or capacity to support similar comprehensive testing of connected and concurrent kill chains. Additionally, participation in Emerald Flag is voluntary and based on interest and availability at the program level—there is no oversight from service leadership or OT&E to coordinate programs or determine the objective of the exercises. It is essential for both program managers and leadership in the testing community to recognize that kill chains do not occur in a vacuum, but as a greater mission-oriented action often across multiple domains.

Connected, concurrent kill chains will increasingly become the norm as cyber-physical systems with new technology are developed. These systems will require test planning requirements to expand beyond those applying to a single program to achieve the required insight into the systems' effectiveness and suitability. Bridging siloed service tests and activities will require that a joint forces approach be developed.

Finding 3-2: Operationally testing connected concurrent kill chains is critically important as the nature of the warfighting environment becomes increasingly complex through the integration of programs and multi-mission systems across multiple domains and the incorporation of advanced technologies with differing degrees and types of human interaction.

Finding 3-3: Testing connected concurrent kill chains drives infrastructure requirements for the ranges that are different from those previously demanded. Ranges require infrastructure that enables seamless and secure communications and data sharing across systems and ranges.

Conclusion 3-2: Testing ranges are not optimized for testing end-to-end kill chains; they were not designed for collaborations with other ranges, and they lack the framework and infrastructure to test concurrent and connected kill chains.

A JOINT PROGRAM OFFICE TO SUPPORT DOD MULTI-DOMAIN TESTING NEEDS

A central feature of test and evaluation in DoD is that it is shaped by program requirements set during their acquisition process. This model works well in tailoring the ranges and range resources to support specific weapon systems or specific technology development but breaks down when cyber physical systems become more prominent and paradigm shifts in technology change the nature of warfare through greater interactions between individual weapon systems and technology.

DoD's Central Test and Evaluation Investment Program (CTEIP), managed by the Test Resource Management Center (TRMC), is DoD's corporate investment program which was established to modernize the DoD test infrastructure. Chapter 5 provides greater detail on test range funding, but it is worthwhile to note that a challenge for the current funding framework is that even if a network of test ranges secures CTEIP funds to build some infrastructure to support multi-domain and kill chain testing, there is no clear funding stream for the sustainment of that joint infrastructure and funding the execution of the testing if that infrastructure does not trace to program needs.[3]

---

[3] Department of Defense (DoD), 2004, "Department of Defense Test Resource Management Center (TRMC)," DoD Directive 26 (DoDD) 5105.71, March 8, https://www.esd.whs.mil/Portals/54/Documents/DD/issuances/dodd/510571p.pdf.

Our nation's superiority in future combat requires the appropriate fielding of programs and systems capable of seamlessly connecting kill chains in a multi-domain environment. The services and program managers overseeing current tests have overlapping and sometimes conflicting requirements and objectives tied to specific programs. Even when there is a recognized need for a test to demonstrate and evaluate the integration of a given system, a large scale test can be cost prohibitive for any one program. Additionally, if a program has funding for an integrated test, the test will likely be bounded to the objectives of the given program and not a full kill chain. There is a need for an environment that is available for programs to participate in to meet their objective but that can also address broader strategic questions related to the integration of multiple systems and technologies across a representative kill chain.

The testing of systems of systems is not new, and a finding of this study is that there are admirable emerging efforts like Orange Flag and Emerald Flag, which have been created to test systems of systems in a multi-domain environment. These have been executed thus far through dedicated efforts of key individuals and funded by pooling resources of participating programs with the objectives of a given exercise shaped by those programs. A concern from this study is how sustainable these efforts are without some dedicated program or office with associated funding to support the sustainment and growth of these capabilities and also how these tests can support not just the program objectives but broader Combatant Command, Joint Staff, or other DoD multi-service objectives. Based on testimony from the public workshop, the services and DOT&E agree that multi-systems testing is critically important, but those tests are ultimately limited by the specific scope of a given program (NASEM, 2021).

There is currently a Joint Test and Evaluation (JT&E) program, but its focus is on concepts of operations for specific use cases that are proposed and approved as stand-alone individual efforts. The primary objective of JT&E is to provide rapid solutions to operational deficiencies identified by the joint military community by developing new tactics, techniques, and procedures (TTPs) and rigorously measuring the extent to which their use improves operational outcomes.[4] JT&E as it is currently organized is not a mechanism to connect mission threads, broad DOT&E test objectives, and the recurring execution of multi-system tests at events like Emerald Flag to address those objectives.

The committee determined that a need exists for a joint program office to enable experimentation and testing of connected concurrently executed kill chains across systems and technologies in a sustained manner to

---

[4] DoD, "FY19 Joint Test and Evaluation (JT&E) Program," https://www.dote.osd.mil/Portals/97/pub/reports/FY2019/other/2019jte.pdf?ver=2020-01-30-115602-597.

assess mission-level capabilities and operational employment. The intent is to provide the means for recurring test events that can provide a "sandbox" that various programs can participate in to meet their program needs and that can address broader strategic objectives. The services and Major Range and Test Facility Base (MRTFB) leadership would still have responsibility of the test execution. This joint program office would include joint service representatives and work with offices in the Joint Staff, Combatant Commands and efforts like the Innovation Steering Group to define key mission threads and information needs related to the integration of domains and emerging technologies in those mission threads. The TRMC and service T&E budgets would still fund the test infrastructure and modernization efforts, and the joint program office would help fund the sustainment of the capabilities for multi-domain tests and the execution of those test events. The program office would work with existing DoD agencies to address cross-service policy and standards that are barriers to these types of tests, to work with Joint Staff on mission threads and with operational and developmental test and evaluation experts on system-of-systems test objectives that will inform test infrastructure requirements, and to identify and provide funding for the execution of those tests that are not part of specific program objectives. This office would include representatives from Joint Staff, combatant commands (COCOMs), the services, the Office of the Secretary of Defense Research and Engineering (OSD (R&E)), and DOT&E.

The committee does not intend to be overly prescriptive concerning the structure of the joint program office because it expects the office will both want and need space to grow and adapt as needed over time, and its success will depend in no small measure on decisions regarding its funding and authorities. A potential location for a joint program office as described above is the Joint Staff J8 Force Structure, Resources, and Assessment Directorate, since COCOMs J8s similarly plan and oversee joint warfighter technology demonstrations. An alternative is that this is an office that falls under the oversight of DOT&E similar to the current JT&E office or it is a growth and expansion of the current JT&E mission. DOT&E currently has the authorities necessary to establish the recommended office.

> **Recommendation 3-1: To enable a range of the future that is capable of testing kill chains and multi-domain operations that can integrate effects across National Defense Strategy modernization areas, the Secretary of Defense should address the need to enable Department of Defense ranges to provide regular venues to "test as we fight" for acquisition and prototyping programs in a joint multi-domain battlespace of integrated systems.**

The committee envisions that this effort would:

a. Reside in DOT&E and report to a committee chaired by the DOT&E and consists of representatives from the Joint Staff, COCOMs, the services, and R&E;
b. Establish clear definitions for "multi-domain operations" and "cyber-physical systems";
c. Lead an effort across Joint Staff elements to define representative multi-domain use cases as well as OT&E objectives and range testing requirements;
d. Work with COCOMs on operational community needs for test information/results to inform operations;
e. Work with technology prototype efforts, e.g., Joint Capability Technology Demonstration, to understand and inform test objectives related to the integration of new technology to enable rapid capability integration;
f. Provide inputs to programs and services on needed future developments based on MDO test results;
g. Provide and advocate for funding to support execution of multi-domain test events and sustainment of capabilities needed to execute those events;
h. Assist with the prioritization of MDO and kill chain tests and associated test resources; and
i. Establish a shared, accessible, and secure modeling and simulation (M&S) and data ecosystem to drive integrated development and testing across the life cycles of multiple supporting programs.

Throughout the remainder of this report specific needs for testing MDOs and connected, concurrent kill chains will arise. The challenge of future testing will be the availability of the minimal set of range capabilities to adequately test the effectiveness and suitability of new systems within this environment. Specific range capabilities, as well as additional enterprise capabilities, are necessary to adequately achieve this minimal set of range capabilities. Box 6.1 in Chapter 6 provides a summary of the necessary range capabilities highlighted throughout this report that are critical for meeting operational testing needs through 2035.

MITIGATING ENCROACHMENT TO SUPPORT FUTURE COMBAT TESTING

Figure 3.2 illustrates the growing size and complexity of the battlespace. As the battlespace becomes larger and more dynamic to encompass interacting systems across multiple domains, the demand for mission space for testing increases. With next-generation weapons that can fly, sail, or drive faster and farther than before, as well as interface with a variety of communication networks, test ranges require more mission space and broad access to the electromagnetic spectrum; however, the testing community is facing a reduction in mission space and a narrowing operating area within the electromagnetic (EM) spectrum.

The DOT&E, the Government Accountability Office (GAO), the TRMC, the Readiness and Environmental Protection Integration (REPI) program, and the MITRE Corporation have all recognized that retaining adequate mission space to meet test requirements is critical (DoD, 2020b; DOT&E, 2020; GAO, 2017; Lachman et al., 2007; MITRE, 2007; TRMC, 2010). The 2018 Sustainable Range Report also highlighted this challenge by noting that "emerging technologies such as hypersonics, autonomous systems, and advanced subsurface systems will require enlarged testing and training footprints." (DoD, 2018b). This concern was echoed by the service test and evaluation executives and program representatives at the public workshop (NASEM, 2021), and representatives from the MDA and NTTR noted that encroachment was a growing concern for conducting operational tests at their locations.

In the context of DoD test ranges, encroachment refers to any factors that obstruct, impede, or suppress the ability of the test community to conduct operational test and training exercises. DoD Directive 3200.15 (DoD, 2013) defines encroachment as "external, as well as internal, DoD factors and influences that constrain or have the potential to inhibit the full access or operational use of the live training and test domain." Encroachment inhibits full access to the live training and test domain by restricting access to the resources necessary to conduct tests. This can be the physical space, or ranges, controlled by the services, which provide the backbone of test area for the test and evaluation community. Because the majority of current and next-generation weapon systems are dependent on the electromagnetic spectrum, encroachment of the electromagnetic spectrum can also inhibit the operational use of these domains (CRS, 2021b).

Encroachment was recognized when GAO reported in 2002 that DoD lacked a comprehensive plan to manage encroachment on ranges that dealt with test and training operations (GAO, 2002). In an attempt to address emerging encroachment concerns, Congress established the Conservation Partnering Program (CPP) and Sustainable Ranges Initiative (SRI) to collaborate with community organizations and provide investments to create exclusion areas around test and training locations. The CPP is currently known as REPI.

A 2007 assessment of the REPI program by the RAND Corporation found that it appeared to be successful to that point but that more could be done to protect DoD mission space (Lachman et al., 2007). The report's recommendations included suggestions that DoD address fundamental causes of encroachment, increase OSD and service investments, and develop additional local partnerships. Following the release of the report, OSD, the services, and DOT&E developed mitigation efforts to address encroachment concerns. A 2016 GAO report outlined these efforts and provided a framework for implementing additional collaborative mechanisms to prevent and mitigate encroachment (GAO, 2016).

With mission space limited, which is a perennial issue for the test community, mission capability will be lost as programs become limited in what they are able to test on a live range. At the committee's January workshop, Conrad Grant, the chief engineer at Johns Hopkins University Applied Physics Laboratory, explained that programs are not able to replicate live end-to-end testing for boost-glide hypersonic vehicles and ballistic missile defense systems (NASEM, 2021).

An example of a recent high-profile encroachment concern occurred at the Eastern Gulf Test and Training Range (EGTTR). Managed by Eglin Air Force Base, EGTTR controls more than 120,000 square miles of airspace and has historically been protected from encroachment under the 2006 Gulf of Mexico Energy and Security Act,[5] which set a moratorium on oil and gas exploration near the EGTTR.[6] This moratorium was set to expire in 2022 until the Trump administration issued a memo extending the moratorium to 2032.[7] While this memo extended the moratorium, it highlights the fragility of the mission space available to DoD. This memo can be reversed at any time, leaving EGTTR vulnerable to the loss of critical mission space, even as plans are being put into place to expand the reach of EGTTR to allow for testing of 5th and 6th generation weapon systems through the Gulf Range Enhancement Program.

Persistent External Encroachment Threats

DoD Directive 3200.15 distinguishes between encroachment caused by external factors and encroachment caused by internal factors (DoD,

---

[5] Gulf of Mexico Energy Security Act of 2006, United States Congress 1331, https://www.boem.gov/sites/default/files/oil-and-gas-energy-program/Energy-Economics/Econ/GOMESA.pdf.

[6] Testimony from Protecting and Securing Florida's Coastline Act of 2019, *Congressional Record*, Volume 165, Number 145, September 11, 2019, https://www.govinfo.gov/content/pkg/CREC-2019-09-11/html/CREC-2019-09-11-pt1-PgH7622.htm.

[7] Presidential Memoranda, "Memorandum on the Withdrawal of Certain Areas of the United States Outer Continental Shelf from Leasing Disposition," September 8, 2020, https://trumpwhitehouse.archives.gov/presidential-actions/memorandum-withdrawal-certain-areas-united-states-outer-continental-shelf-leasing-disposition/.

2013). Figure 3.3 illustrates the external encroachment threats identified by the REPI program from fiscal year 2020 (DoD, 2020b). Encroachment concerns have continued to rise in recent years, with noise complaints, residential and commercial growth, environmental impacts, and spectrum use the leading causes for concern (DoD, 2018b). Land encroachment, like construction projects, can bring residential areas closer to DoD ranges and threaten testing operations. For example, a rifle range at Camp Butner in North Carolina was shut down due to encroachment from noise complaints, and it is believed helicopter and other training operations will soon be restricted by further noise complaints (DoD, 2020b). Additionally, windmill farms can adversely impact military activies by interfering with air defense radars and increasing ambient seismic noise levels (DoD, 2006). Additional areas of external encroachment include peer and near-peer surveillance using drones, satellites, and other equipment as well as commercial customers using range telemetry and altimeter resources.

A recent example of external encroachment comes from NTTR. The Desert National Wildlife Range (DNWR) placed restrictions on operating areas for NTTR activities in 2016. The affected area, primarily on the south range, is used by NTTR to conduct flight testing, classified research and development projects, and weapons tests (Aftergood, 2020). A 2017 proposal that would have expanded protected areas for NTTR to operate was

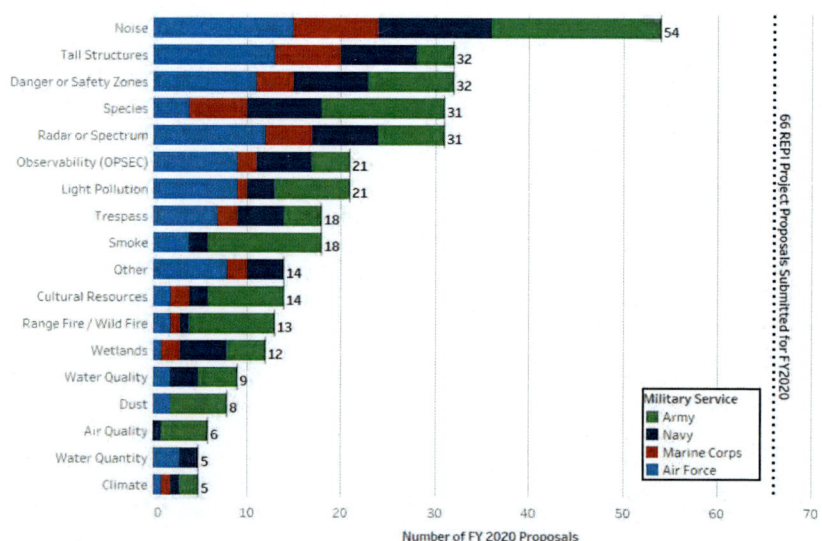

FIGURE 3.3 Encroachment threats identified in fiscal year (FY) 2020 Readiness and Environmental Protection Initiative (REPI) proposals. SOURCE: DoD (2020b).

denied.[8] Often referred to as the NTTR Land Withdrawal Strategy, this proposal was intended to protect the land needed to conduct test operations. Representatives of NTTR expressed concern during the committee site visit that the rejection of this strategy will directly lead to the loss of mission capability (see Appendix B).

Naval test ranges face encroachment from windmill farms off the Virginia and North Carolina coasts, which threaten to infringe on the already limited available space on Atlantic test and training ranges (Niiler, 2019). Naval Base Kitsap also suffered from noise encroachment due to increased acoustic interference from surrounding vessels, forcing a response from the REPI program to ensure that missions at Naval Base Kitsap could continue (DoD, 2020b). Windmills also interfere with radars used in testing processes (DOE, 2018).

Electromagnetic Spectrum Encroachment

External encroachment also includes the declining access to various bands of the electromagnetic spectrum. Spectrum encroachment is not a new issue. A 2007 MITRE report identified a "crisis" of insufficient spectrum for the flight test community, which was affecting aeronautical telemetry for the transmission of real-time data during flight tests. Since then, a number of efforts have been undertaken to limit the sell-off or sharing of EM bands deemed critical to the T&E community for testing (MITRE, 2007). However, emerging technologies have complicated this mitigation process. The Radio Technical Commission for Aeronautics released a report in October 2020 recognizing that 5G transmitters cause interference with the radar altimeters used for commercial and military aircraft even though 5G has its own unique bands of operation separate from those used for radar altimeters (RTCA, 2020). This interference directly hinders the ability of the U.S. Air Force to conduct end-to-end system testing in a live setting.

During a committee site visit, representatives from NTTR said that they recognized that spectrum issues currently exist and that they expect them to become even more pressing in coming years. They noted that NTTR no longer receives requests for GPS jamming because they cannot obtain approval from commercial and other government agencies to conduct jamming tests.

The external encroachment from the loss of spectrum directly affects the ability to validate system performance against threat systems. As opposed to telemetry data use of the spectrum, DoD does not have control

---

[8] "Proposal to Withdrawal and Reservations of Public Lands in Nevada to Support Military Readiness and Security," https://fas.org/man/eprint/ndaa-2021-prop/04172020-nevada.pdf.

over where the threat frequencies of U.S. adversaries will operate in operational situations. As a result, frequency sell-off in certain wavelengths removes the ability to conduct operational testing against those threats in a live environment. In addition, the loss of spectrum will lead to fielding systems that have not undergone extensive testing in the EW and other spectrum-related arenas.

> **Finding 3-4:** As frequency is sold off, the test community loses the ability to conduct operational testing in live environments against certain threats.

> **Conclusion 3-3:** Encroachment leads to the inability to demonstrate mission capability and identify deficiencies due to lack of access to the physical and electromagnetic spectrum space with which to conduct test and evaluation. This creates operational risk as DoD will have to field weapon systems that have not been tested against certain threats.

The committee recognizes that spectrum sharing for DoD is a necessity. This necessity, however, does not preclude the need for bands to be reserved solely for DOT&E testing of systems that require exclusive access to certain bands within the spectrum. Adopting a successful spectrum management strategy is critical to retaining control of the electromagnetic spectrum to ensure that systems have the required frequencies available for proper operation.

The development of a spectrum management strategy could be initiated by DOT&E by conducting a review and identifying critical bands within the spectrum that are necessary for the operational testing of next-generation weapon systems at live ranges. DOT&E would then collaborate with stakeholders and recommend action to protect those critical bands so that live testing of EW and other weapons will be able to take place on next-generation weapon systems. While previous studies have focused on spectrum loss from a telemetry data perspective, the following recommendation is focused on the need to identify operational risk and impact of threats that cannot be tested against in a live environment due to spectrum sell-off:

> **Recommendation 3-2:** To ensure the ability to validate the survivability of Department of Defense (DoD) weapon systems against a realistic operational threat environment across air, sea, land, space, and cyberspace domains, DoD should identify and prioritize bands that cover U.S. military operational and test requirements which should be protected from sell-off to preserve these capabilities.

Recommendation 3-3: The Test Resource Management Center (TRMC) should assess current and projected commercial radio frequency communications technologies and spectrum allocations for secure, agile, high-bandwidth operational test needs. In addition, TRMC should determine the feasibility of developing new large-scale enclosed testing facilities combined with expanded modeling and simulation to support electromagnetic spectrum activities not suitable for open-air testing.

Internal Encroachment Challenges

Internal encroachment refers to actions taken by DoD that result in encroachment. Increased demand and tempo at ranges can lead to actions that restrict the ability of test groups to perform the full range of necessary tests. An example of this is when the 7th Special Forces Group was moved to Eglin AFB as a result of the 2005 Defense Base Closure and Realignment Commission, which recommended the move as "an opportunity to achieve outstanding joint training through its collocation with the Air Force Special Operations Command" (DBCRC, 2005). Personnel from the Eglin/Edwards/Wright Patterson Air Force bases site visit discussed how a new compound for the 7th SOG (Studies and Observations Group) was placed into the center of test space at Eglin Air Force Base. This action directly resulted in the cancellation of 29 operational test profiles, including major test programs like the F-16, Small Diameter Bomb test program, Joint Standoff Weapon, and several other classified test programs. There are examples where the movement of range equipment (both radiating and non-radiating), uncoordinated with test operations have resulted in test anomalies and "no test" results, causing significant unplanned analyses and the unnecessary consumption of critical test assets that have delayed the Initial Operating Capability (IOC) of major systems by months or even years.

As evidenced by the increase in encroachment concerns since 2001, external encroachment from business, residential, and foreign entities will be a constant challenge in the future and will require mitigation strategies spanning from the development of bladeless wind turbines to land preservation negotiations to mapping noise corridors (GAO, 2016). These strategies, however, are either already being pursued by programs such as REPI or are outside the scope of this study. While the committee recognizes the existence of external encroachment issues, efforts are under way to create frameworks and recommendations for mitigating these concerns.

There is, however, a lack of literature on internal encroachment issues and how they might be mitigated. By definition, internal encroachment is caused by DoD and thus offers DoD an opportunity to limit the impacts of encroachment in coming years by restricting actions taken within the department to limit the space available for test and training operations.

Finding 3-5: External encroachment will continue to be a persistent threat for DoD, but issues of internal encroachment, if left unaddressed, will cause unintended consequences for MDO T&E.

Given that there already exists a program to identify and mitigate external encroachment issues facing the test ranges, rather than suggest the establishment of a new program, the committee recommends the following:

Recommendation 3-4: The Department of Defense should broaden the authority of the Test Resource Management Center (TRMC) to address issues of internal encroachment by reviewing internal range policies and actions to ensure that the test groups retain adequate mission space and prevent the placement of equipment or infrastructure that could potentially interfere with test operations. The Director of Defense Research and Engineering for Advanced Capabilities should be granted the authority to mitigate disputes arising over internal encroachment concerns and provided additional funding to manage internal encroachment.

Encroachment Challenges for Next-Generation Systems

As next-generation weapon systems enter OT&E, their onboard systems and sensors, as well as the ability for multiple platforms to integrate and act collectively, exceed the capabilities of current ranges and the existing range capabilities constrain the ability to test these advanced systems. In the case of hypersonic weapons, the significant increase in sustained speed, distance, and impact of these weapons coupled with the number of programs pursuing this technology result in tests that exceed the capabilities of historic approaches and test locations. These constraints were voiced to the committee at their March 4, 2021, meeting by Michael White, the principal director for hypersonics, who used the phrase "string of pearls" to describe how testing will have to be conducted going forward. "String of pearls" refers to the integration of multiple ranges together in a single hypersonic test in order to track the hypersonic vehicle throughout the entire trajectory of the flight (Spravka and Jorris, 2015). This string-of-pearls issue also faces space testing, long-range precision fires, and intercontinental ballistic missile (ICBM) operational testing.

Spectrum encroachment issues will become more pronounced as next-generation weapon systems enter OT&E. Any further restriction on spectrum access will "directly impact DoD's ability to conduct live training" (DoD, 2018b). EW systems, for example, are classified as spectrum dependent systems (SDS) which require the use and control of spectrum resources. DOT&E testing generally contains requirements for systems to test using spectrum in the live environment for either electronic attack (EA), electronic protection (EP), or electronic warfare support (EWS). The loss of spectrum has resulted in nonrealistic training scenarios and limited the ability to execute TTPs.

> **Finding 3-6:** The DoD has taken actions to preserve mission space in recent years, but the performance of critical systems exceeds the boundaries of current ranges. This problem will become increasingly worse as advanced multi-domain operations further stress available test facilities.

Given the existing encroachment issues facing U.S. military ranges, the growing need for adequate physical and spectrum space in which to conduct tests, and the difficulty of expanding the physical and spectrum boundaries of ranges within the United States, one potential approach would be to cooperate with foreign allies to invest in additional test range space. As an example, Australia in particular has physical space that is expansive enough for hypersonic weapon testing, and a collaborative agreement could allow both countries to test new military technologies and scenarios over much larger areas than currently available to the United States. These efforts are permissible through 10 U.S. Code §2350l, which gives authority to the Secretary of Defense to enter an agreement with a foreign country to provide testing of U.S. defense equipment at that country's test facilities.[9] Military testing cooperations with foreign countries are overseen by the International Test and Evaluation Program (ITEP), which is managed by DOT&E. Given TRMC's authority to maintain awareness of testing needs for current and future technologies, they are suitable for determining the use and investment strategies for testing military systems abroad.

> **Recommendation 3-5:** The Test Resources Management Center should develop a strategy that assesses the use of and potential investment in suitable allied resources for open-air testing. This strategy should include criteria for the usage of allied resources and areas of potential investment to include range space available, data collection, security risks, and support facilities.

---

[9] United States Code, 2012 Edition, Supplement 2, Title 10 - ARMED FORCES. https://www.govinfo.gov/content/pkg/USCODE-2014-title10/pdf/USCODE-2014-title10.pdf. Accessed June 22, 2021.

REFERENCES

Aftergood, S. 2020. "Air Force Calls for Expansion of Nevada Test Range." June 1. Federation of American Scientists. https://fas.org/blogs/secrecy/2020/06/nttr-expand/.

American Battlefield Trust. n.d. "Civil War Ballooning." https://www.battlefields.org/learn/articles/civil-war-ballooning. Accessed March 27, 2021.

Brose, C. 2020. *The Kill Chain: Defending America in the Future of High-Tech Warfare*. New York, NY: Hachette Books.

CRS (Congressional Research Service). 2021a. "Defense Primer: Army Multi-Domain Operations (MDO)." October 22. Washington, DC. https://fas.org/sgp/crs/natsec/IF11409.pdf.

CRS. 2021b. *Overview of Department of Defense Use of the Electromagnetic Spectrum*. August 10. Washington, DC. https://fas.org/sgp/crs/natsec/R46564.pdf.

DBCRC (Defense Base Closure and Realignment Commission). 2005. *2005 Defense Base Closure and Realignment Commission Report to the President*. https://www.acq.osd.mil/brac/docs/BRAC-2005-Commission-Report.pdf.

DoD (Department of Defense). 2006. *The Effect of Windmill Farms on Military Readiness*. https://archive.defense.gov/pubs/pdfs/WindFarmReport.pdf.

DoD. 2013. *Directive 3200.15*. Washington, DC. https://www.esd.whs.mil/Portals/54/Documents/DD/issuances/dodd/320015p.pdf.

DoD. 2018a. *Summary of the 2018 National Defense Strategy of the United States of America*. https://dod.defense.gov/Portals/1/Documents/pubs/2018-National-Defense-Strategy-Summary.pdf.

DoD. 2018b. *2018 Report to Congress on Sustainable Ranges*. Washington, DC.

DoD. 2020a. *2020 Report to Congress on Sustainable Ranges*. Washington, DC.

DoD. 2020b. *2020 Report on REPI Program Outcomes and Benefits to Military Mission Capabilities*. https://www.repi.mil/Portals/44/Documents/Metrics_Reports/2020_REPI_Metrics_Report_FINAL_LOWRES_29SEP20.pdf.

DOE (Department of Energy). 2018. *Wind Turbine Radar Interference Mitigation*. https://www.energy.gov/sites/prod/files/2018/04/f51/WTRM_Factsheet_Final_2018.pdf.

DOT&E (Director, Operational Test and Evaluation). 2020. *FY 2020 Annual Report*. https://www.dote.osd.mil/Publications/Annual-Reports/2020-Annual-Report/.

GAO (Government Accountability Office). 2002. *Military Training: DoD Needs a Comprehensive Plan to Manage Encroachment on Training Ranges*. Washington, DC. https://www.gao.gov/assets/gao-02-727t.

GAO. 2016. *Defense Infrastructure: DoD Efforts to Prevent and Mitigate Encroachment at its Installations*. Washington, DC. https://www.gao.gov/assets/gao-17-86.pdf.

GAO. 2017. *Military Training: DoD Met Annual Reporting Requirements in its 2017 Sustainable Ranges Report*. Washington, DC. https://www.gao.gov/assets/690/688137.pdf.

Grest, H., and H. Heren. 2019. *What is a Multi-Domain Operation?* Joint Air Power Competence Centre. https://www.japcc.org/what-is-a-multi-domain-operation/.

Lachman, B.E., A. Wong, and S.A. Resetar. 2007. *The Thin Green Line: An Assessment of DoD's Readiness and Environmental Protection Initiative to Buffer Installation Encroachment*. Arlington, VA. The RAND Corporation. https://www.rand.org/pubs/monographs/MG612.html.

MITRE. 2007. *The Economic Importance of Adequate Aeronautical Telemetry Spectrum*. McLean, VA. https://www.mitre.org/sites/default/files/pdf/07_0187.pdf.

NASEM (National Academies of Sciences, Engineering, and Medicine). 2018. *Multi-Domain Command and Control: Proceedings of a Workshop—In Brief*. Washington, DC: The National Academies Press. https://doi.org/10.17226/25316.

NASEM. 2021. *Key Challenges for Effective Testing and Evaluation Across Department of Defense Ranges: Proceedings of a Workshop–In Brief.* Washington, DC: The National Academies Press. https://doi.org/10.17226/26150.

NATO (North Atlantic Treaty Organization). 2016. AJP-3.9 "Allied Joint Doctrine for Joint Targeting." https://assets.publishing.service.gov.uk/government/uploads/system/uploads/attachment_data/file/628215/20160505-nato_targeting_ajp_3_9.pdf.

Niiler, E. 2019. "The Military Is Locked in a Power Struggle with Wind Farms." *Wired.* May 20. https://www.wired.com/story/the-military-is-locked-in-a-power-struggle-with-wind-farms/.

RTCA (Radio Technical Commission for Aeronautics). 2020. *Assessment of C-Band Mobile Telecommunications Interference Impact on Low Range Radar Altimeter Operations.* Washington, DC. https://www.rtca.org/wp-content/uploads/2020/10/SC-239-5G-Interference-Assessment-Report_274-20-PMC-2073_accepted_changes.pdf.

Spravka, J.J., and T.R. Jorris. 2015. *Current Hypersonic and Space Vehicle Flight Test and Instrumentation.* Edwards Air Force Base, California. https://apps.dtic.mil/sti/pdfs/ADA619521.pdf.

TRMC (Test Resource Management Center). 2010. *FY2010 Annual Report.* Washington, DC.

Tirpak, J.A. 2000. "Find, Fix, Track, Target, Engage, Assess." *Air Force Magazine.* https://www.airforcemag.com/article/0700find/.

4

Digital Infrastructure Needs for Operational Testing

For decades, the power and speed of, and connectivity afforded by, digital technologies have been increasing exponentially, with implications for every part of society. Not surprisingly, digital technologies are also dramatically reshaping both military technologies and the ways in which those technologies are developed, tested, and deployed. This will pose increasingly serious challenges to the military's operational testing and evaluation (OT&E) over the coming years while simultaneously offering opportunities to make OT&E more responsive, effective, and flexible.

This chapter examines those two complementary aspects of digital technologies in operational testing—the challenges and the opportunities. Many of the challenges arise because of the appearance of novel military technologies whose operational testing requires approaches that are fundamentally different from anything in existence today. Perhaps the best example of this is found in the areas of artificial intelligence, autonomous systems, and machine learning, which are posing new and—so far—unresolved challenges for those who seek to test the operational performance of such technologies.

At the same time, digital technologies are providing new and powerful approaches to OT&E. One example is the rise of digital twins and high-performance modeling and simulation, which are enabling novel ways of testing technologies and systems. David Tremper, Director of Electronic Warfare for the Office of the Secretary of Defense, addressed the committee at the public workshop and shared successes realized in the AEGIS combat system's use of an onboard digital twin that can operate simultaneously with the operational system. The appearance of digital twins is particularly timely, given that the combination of new domains and operational constraints is increasingly making virtual testing the only practical approach for certain applications.

Finally, as ever more powerful digital technologies are enabling the collection, processing, and analysis of massive amounts of data from testing, military ranges will be increasingly challenged to collect, process, transmit, store, and analyze these data securely and effectively. An additional challenge will be securing these data in all of their states and ensuring the data is accessible, secure, and consumable to those who need it since data generated during operational testing may be at a mix of classification levels. These combined challenges are placing growing demands on the digital infrastructure of the nation's military ranges.

MODELING AND SIMULATION

Modeling and simulation (M&S) is becoming an increasingly essential part of operational testing. This growing role is driven by a number of factors, some on the supply side and some on the demand side. On the supply side, rapid increases in computing power and memory combined with improvements in software capabilities and sophistication have dramatically expanded what is possible to do with digital models. M&S now plays a major role in the development of commercial products, such as automobiles (Biesinger et al., 2019) or pharmaceuticals (USFDA, 2021), dramatically shortening the time it takes to bring a product to market, and it has the potential to dramatically improve the testing of military weapons and systems as well. A case study in the effectiveness of this approach is the National Nuclear Security Administration's Science Based Stockpile Stewardship (Reis et al., 2016) program in which high performance computer simulations across multiple physics and materials science disciplines play a leading role in certifying the safety and security of nation's nuclear stockpile. In fact, the Deparment of Defense (DoD) is already building capabilities to use M&S to support future technologies such as advanced aircraft,[1] space systems,[2] and artificial intelligence.[3]

On the demand side, a variety of factors are driving the growing role of M&S in OT&E. For example, some test exercises would reveal sensitive information and capabilities. Since it is unrealistic to hide open-air tests

---

[1] Panel discussion from John Pearson, Senior Evaluator, 5th/6th Generation Fighter Aircraft to Workshop on Assessing the Suitability of Department of Defense Ranges, January 28, 2021.

[2] Panel discussion from COL Eric Felt, Director, AFRL Space Vehicles Directorate to Workshop on Assessing the Suitability of Department of Defense Ranges, January 28, 2021.

[3] Panel discussion from Brian Nowotny, DoD Autonomy Test Lead, Test Resource Management Center to Workshop on Assessing the Suitability of Department of Defense Ranges, January 29, 2021.

or space-based tests from observation, any such testing risks providing information to U.S. adversaries about the capabilities of the systems being tested. If certain details need to remain secret, testing via simulation is often the best, or only, option.[4]

Another reason to simulate is that some systems simply cannot practically be tested across their entire application space. For example, open-air testing of hypersonic weapon systems requires large geographical areas at various altitudes, but these systems are expensive single-use devices that are too costly and time prohibitive to fully explore the operational envelope in test.

It would also not be feasible to carry out full-scale tests in a real-world environment of a weapon designed to compromise nearby computers and digital communications. Furthermore, some testing environments cannot be physically replicated on DoD test ranges. For example, artificial intelligence (AI) systems must train and execute on a stream of operational data that reflects their intended operating environment. During development and testing, an autonomous vehicle has access to the same roads as the operational system. However, an AI system to classify threat emitters cannot have constant access to the electromagnetic emissions of anticipated future threats. On the other hand, simulation models, informed and improved by intelligence over time, may be run to generate sample data.

Finally, running simulations is generally less expensive than running tests with expensive pieces of equipment, and while simulations cannot completely replace physical tests—real-world data will always be necessary for grounding models in reality—simulations can be used in various ways in conjunction with testing, and should be embedded in the planning of test programs.

Traditional Use of Modeling and Simulation in Weapons Testing

The classical view of the role of M&S in testing is illustrated in Figures 4.1 and 4.2, taken from a Defense Acquisition University (DAU) training class on the role of M&S in testing (DAU, n.d.). Figure 4.1 illustrates a classical linear process where a test is developed, results are predicted by the M&S tools, the test is executed, and the results are compared to the predictions. This approach has proven successful in accelerating the pace of testing in many programs and has demonstrated some of the promise of M&S in system level testing. However, this view does not fully exploit the power of modern M&S.

---

[4] Panel discussion from Dr. Raymond D. O'Toole, Acting Director, Operational Test and Evaluation to Workshop on Assessing the Suitability of Department of Defense Ranges, January 28, 2021.

DIGITAL INFRASTRUCTURE NEEDS FOR OPERATIONAL TESTING

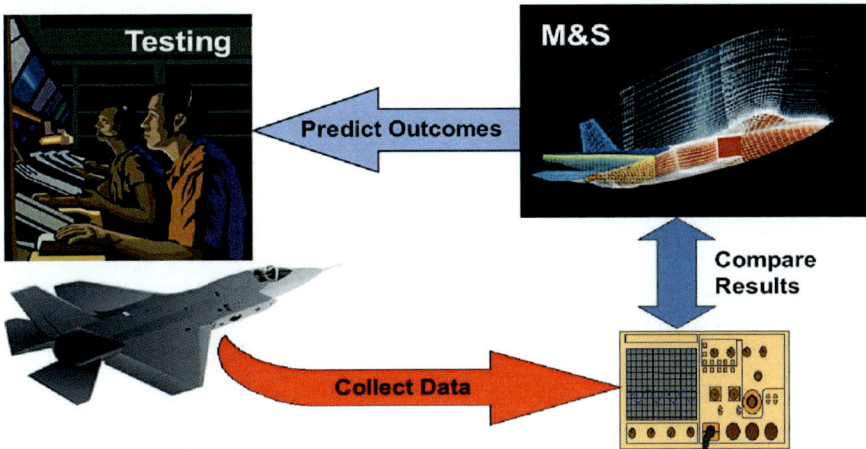

FIGURE 4.1 Classical view of the role of modeling and simulation (M&S) in system-level testing. SOURCE: DAU (n.d.).

This classical view is expanded in Figure 4.2, which highlights the major role of M&S at the program outset in supporting design activities, but with that role diminishing over the program life cycle. Under this approach, the M&S tools developed early in the program are frequently not sustained or evolved to perform an integrated function with developmental testing (DT) and operational testing (OT). Too frequently, once the system enters test, models are redeveloped from scratch, without good linkage to the models that were employed early in the program. Another result of this approach is that the system will frequently be turned over to the warfighter without prior exposure to robust system-level models.

Benefits of Modeling and Simulation in Testing

Simulation should not be viewed only as a replacement for testing or as an alternative that is less expensive or more convenient or that can be carried out in situations where physical testing cannot. Rather, simulation is a fundamentally different approach to testing systems that has its own benefits and advantages that are different from and complement those of physical testing. This means, in particular, that a thoughtful combination of simulation with testing can be much more powerful and effective than either simulation or testing alone. These benefits are presented below:

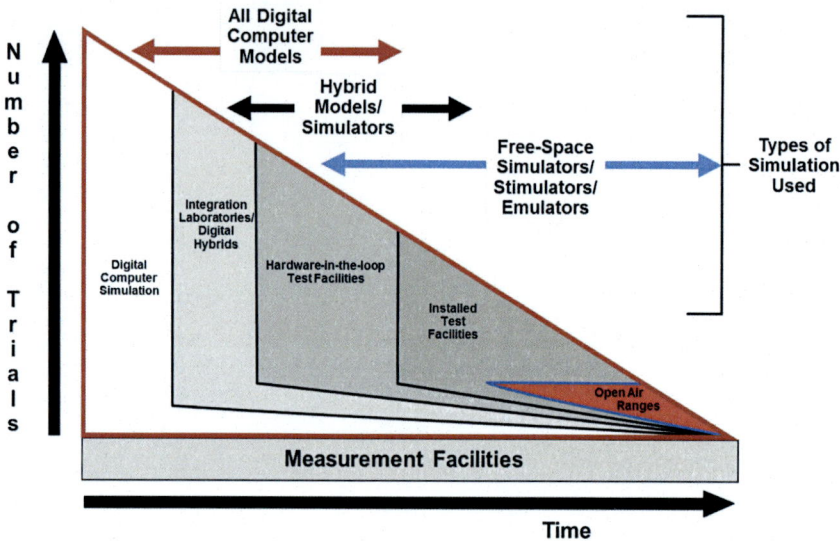

FIGURE 4.2 Classical view of the role of modeling and simulation (M&S) throughout the development life cycle. SOURCE: DAU (n.d.).

1. **A robust simulation environment provides understanding that physical testing cannot.** Physical testing will likely be considered the gold standard in operational testing for some time to come. It is important to recognize, though, that testing largely provides a binary result: either the system worked or the system failed in this one test. By contrast, the simulation environment can help a user understand the system's margin in successful tests and identify—through Monte Carlo analysis, for example—those systems that might have been on the edge of failure. Thus, integrating simulation with physical testing will yield a richer characterization of the system and its performance for the warfighter.
2. **Embracing simulation will drive some testing needs.** For a simulation to be useful, the community must have confidence in the validity of the results. Fortunately, a well-formed simulation environment makes it possible to study the sensitivity of the results to individual parameters in the model and to quantify the uncertainty of the overall result. The discipline of uncertainty quantification has matured to the point that it is possible both to understand the uncertainty in a simulation result and to understand what parameters drive that uncertainty. Testing can then be focused in a manner that maximizes its value in terms of driving down the overall uncertainty in a simulation and that

consequently improves the understanding of the system. Note that the testing required to reduce uncertainty will include both ground testing and flight testing.

3. **Increased use of simulation at the program level can create opportunities at the campaign level.** The major challenges of modern warfare include optimizing and assessing the performance of strike packages that integrate advanced technology systems with legacy systems or multiple advanced technology systems into a combined force. As individual programs establish more sophisticated simulations, the opportunity exists to conduct simulations at the campaign level. Such campaign-level simulations can provide powerful insights—revolutionary technologies can enable revolutionary operational strategies.

4. **The increased use of simulation is unlikely to reduce the load on test ranges.** It is unrealistic to think that simulation will dramatically reduce the need for testing advanced technology systems. The flexibility offered by emerging systems increases the challenges in operational testing in demonstrating that the systems function properly across their entire potential application space. A well-formed operational testing program will integrate simulation with live testing to maximize the demonstrated capability of the system. For many of the advanced technology systems, the threshold testing requirements to develop good confidence in the system will be substantial.

Taking Full Advantage of the Power of Simulation

The current OT&E enterprise has the opportunity to significantly broaden its use of M&S. In short, the increasing power of computers combined with the growing sophistication and effectiveness of digital models is opening up new possibilities in simulation that should be taken advantage of by the nation's military ranges. To use M&S to greatest effect, it will be necessary to integrate testing and simulation more closely than is currently the case.

New Abilities in Computing Are Opening Up New Possibilities in Modeling and Simulation

Advances in modeling and simulation combined with high-performance computing now provide a powerful capability to employ physics models to understand the performance of advanced-technology weapon systems. Modern M&S tools can provide high-fidelity predictions of the behavior of systems under test with reasonable computing times. Combined with approaches such as Monte Carlo analysis and uncertainty quantification, modern simulation capabilities can provide powerful insights into both the performance margin of the system under test and the sources of uncertainty in the behavior of that system. Cloud computing, virtualization, continuous integration/continuous delivery (CI/CD), and DevSecOps approaches allow simulation developers to provide simulation capabilities as a service (Siegfried, 2021), with ubiquitous access to simulation software and on-demand computational power. Such M&S capability is not only valuable in the design phases of a new system, but it can continue to be evolved to support an integrated role in both DT and OT. In addition, such models have sustained value in supporting both effectiveness assessment and eventual campaign level simulations.

The Importance of Integrating Development and Testing with Simulation

An opportunity exists to further accelerate the pace of testing and to improve understanding of the performance of advanced technology systems by embracing a more integrated approach to development, testing, and simulation. In this approach, illustrated in Figure 4.3, DoD develops and sustains a persistent M&S and data environment for a particular military domain. As shown in the figure, a persistent M&S and data environment supports all phases of the life cycle. In the early phase, M&S supports concept development and digital engineering to inform requirements and rapidly assess the effectiveness of engineering choices. As system development begins, M&S supports developmental testing to verify the performance of components and subsystems. Finally, M&S supports operational testing to validate the system itself using realistic models of its operating environment throughout deployment to assess emerging operational uses as well as the effects of aging on system performance. Furthermore, M&S data and results are shared by different levels in M&S abstraction. For example, component and subsystem models are not necessarily re-used in system models, but their data and results enable the development of more abstract models suitable for system validation. Finally, M&S is not simply used to predict test results; it is used to focus testing in particular areas with the goal of driving down uncertainties in system-level understanding. Under this approach, M&S is sustained and evolves through the program life cycle to support the system throughout its life cycle deployment.

It is important to note that the traditional approach of saddling single acquisition programs with stand-alone M&S capabilities will not work for mission-driven testing. Mission-driven testing typically spans multiple programs. Furthermore, if DoD waits until program initiation to build a

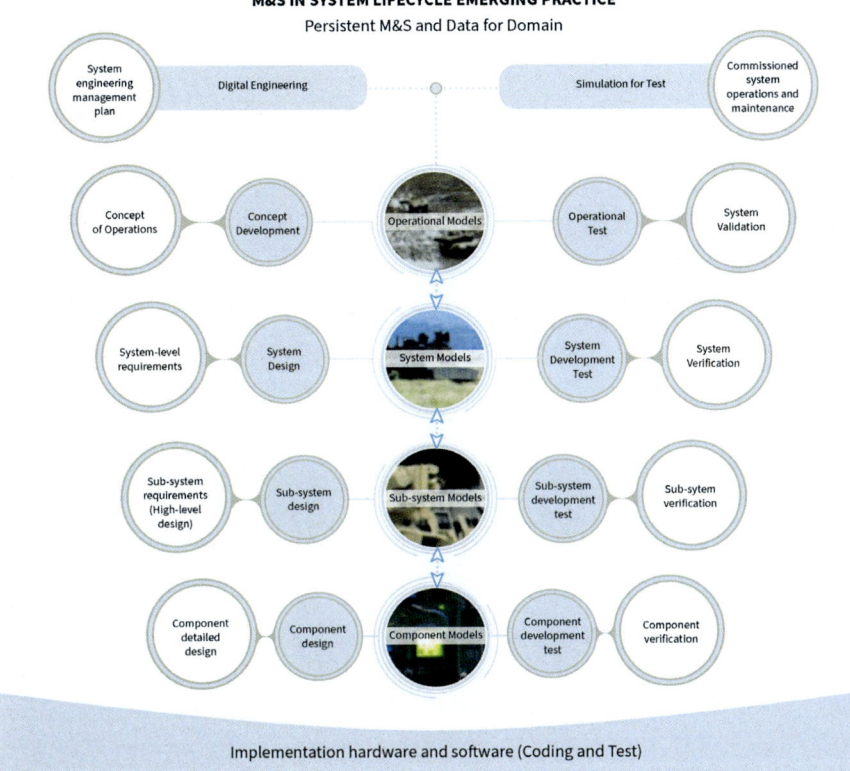

FIGURE 4.3 New paradigm for integrating testing with simulation. SOURCE: Based on image from MITRE Systems Engineering Guide, https://www.mitre.org/publications/systems-engineering-guide/se-lifecycle-building-blocks/test-and-evaluation/verification-and-validation. Images from D. Vergun, "Soldiers' situational awareness improved using micro-displays, augmented reality," Army News Service, March 23, 2018, https://www.army.mil/article/202557; A. Givens, "Army Rolls Out Latest Version of Iconic Abrams Main Battle Tank," U.S. Army, October 9, 2017, https://www.army.mil/article/194952; A. Brutus, "M1A2 Abrams tank live-fire Bulgaria," U.S. Army, June 25, 2015, https://www.army.mil/article/151190; U.S. Army Europe and Africa, "1st Cavalry Division: Multinational Combined Arms Live Fire Exercise," photo, February 11, 2020, https://www.europeafrica.army.mil/Newsroom/Photos/igphoto/2002251824/.

test capability, it will not be available for digital engineering during the early life-cycle phases. Instead, a separate office must develop requirements for, fund, and sustain persistent M&S for critical warfighting domains.

The Joint Simulation Environment (JSE), originally built for F-35 Joint Strike Fighter (JSF) testing in a simulated environment, is an illustrative example. During his opening comments to the committee, Robert Behler, the Former Director of Operational Test and Evaluation, discussed the role of the JSE in evaluating the ability to perform high-level missions against near-peer threats,[5] and he recommended a site visit to Patuxent River. For context, the JSE has had some challenges, and its delayed development has delayed the F-35 initial operational test and evaluation (IOT&E) and full rate production decision because of the daunting challenges of integrating the necessary threat simulators and friendly supporting system simulators in sufficient numbers to represent the theater of interest.[6] However, analysis has shown that overcoming these challenges is feasible.[7]

During their site visit to Air Combat Environment Test & Evaluation Facility (ACETEF), committee members found the JSE to be an innovative and modern approach to simulation of complex DoD mission threads which could be replicated for other missions. However, the program provides some lessons learned. JSF missions typically involve several other DoD systems, so accurate representation of those external systems is difficult for a separately funded program to develop and manage. The JSE is still in development, even though the JSF is well into production, so the JSE was not available to support early life-cycle digital engineering. Additionally, multiple programs could utilize and benefit from the threat models and the theater representation built for JSE, but there is no clear mechanism to make the models accessible and consumable by other programs.

In considering the approach illustrated in Figure 4.3, it is important to keep in mind that programs do not need to decide between testing and simulation. The issue is often framed in terms of the question "Can

---

[5] Robert Behler, Director of Operational Test and Evaluation, "Study Sponsor Perspective," to Kickoff Meeting for Assessing the Suitability of Department of Defense Ranges, December 4, 2020.

[6] Lieutenant General Eric Fick, Program Executive Officer, "Update on F-35 Program Accomplishments, Issues, and Risks," F-35 Joint Program Office Testimony to House Armed Services Committee, Subcommittees on Tactical Air and Land Forces and Readiness Joint Hearing, April 22-23, 2021, https://armedservices.house.gov/2021/4/subcommittees-on-tactical-air-and-land-forces-and-readiness-joint-hearing-update-on-f-35-program-accomplishments-issues-and-risks.

[7] *Defense Daily*, "Lack of F-35 Full-Rate Production Decision Provides 'Launching Point for Criticism of Program,' PEO Says," May 13, 2021.

simulation be used to replace testing (and reduce cost)?" when the better framing is to ask "How can simulation be used *with* testing to maximize the effectiveness of advanced technology systems for the warfighter?" The optimal strategy may involve integrating testing and simulation in a way that produces the greatest value, maximizing understanding and minimizing the cost for testing the program.

A Vision for the Future of Modeling and Simulation in OT&E

Modeling and simulation have tremendous potential to transform OT&E into a much faster, more powerful, and more cost-effective enterprise, but the overall effectiveness of M&S will depend on the details of its implementation. With this in mind, the committee offers the following recommendations for the use of modeling and simulation in support of the nation's military ranges.

Sustain Modeling and Simulation Throughout the Program Life Cycle

The role that M&S will play in the development of a system should be a consideration during Technology Maturation and Risk Reduction (TMRR) and be well matured prior to entry into Engineering and Manufacturing Development (EMD), since it is a key driver to the associated activities, schedules, and cost. To effectively apply M&S for design decisions, system integration, and verification, the detailed requirements for specific models and simulations must be provided early to the M&S community and embedded in program milestones. Delivery of defined modeling capability in terms of functionality, fidelity, and maturity frequently paces system development and will have increasing importance in meeting schedules.

A significant challenge with using M&S and data in program development is that the development of the data and the initiation of M&S often come late in the program so they are only available for system test. If M&S and data resources for a particular set of mission threads were persistent, architected and curated for sharing and joint use, and funded independently of a particular program, they would be available to support many related programs throughout their life cycles. This is why Recommendation 3-1 for a joint program office also includes responsibility to sustain the M&S and data ecosystem necessary for integrated development and test of critical mission threads.

Since M&S has traditionally been used to demonstrate that a preliminary design meets requirements, the associated funding has often been applied well past program inception. However, the most effective use of M&S arises from a well-planned and well-architected infrastructure that provides defined capabilities that are available and scheduled to support early feasibility studies, inform design decisions for multi-disciplinary design optimization, complement integration and test, optimize military operations, support predictable production, guide system sustainment, and realize the benefits of the potential reuse and sharing of models, simulations, analysis tools, and data. This strategy requires a funding profile that is primarily front-loaded and that drives requirements to the M&S community, akin to requirements for prime hardware and software, in order to ensure that well-planned models and simulations are available as scheduled to meet critical program milestones.

The M&S and data ecosystem must be designed so that they support the exchange of attributes among models running in various native applications, with metadata that defines those attributes as well as the functionality, fidelity, and pedigree for each model. The models will evolve over the course of the program and be adaptable and extensible for use at different levels within DoD, providing the required capability at each.

Create a Central Modeling and Simulation Resource

There currently exists service-level offices to support modeling and simulations efforts, such as the Air Force Simulation and Analysis Facility (SIMAF),[8] the Air Force Agency for Modeling and Simulation (AFAMS),[9] and the Army Modeling and Simulation Office (AMSO).[10] However, without a DoD infrastructure to provide common and maintained M&S, each program develops or acquires its own models and emulators to support integration and test against various system interfaces and threats, including command-and-control systems, cues from multiple sources, threat radars, targets, etc. This is inefficient and will likely result in system deficiencies from the use of outdated or inaccurate model representations, which are often developed by organizations that may not have expertise or full insight into those systems. There must be a single, managed, trustworthy source for common DoD models that is maintained as part of a core DoD infrastructure. One particular aspect of these centralized M&S resources should be a collection of digital twins that represent adversary equipment or threats against which multiple development systems will be tested.

M&S also plays an increasingly important role in meeting the challenges associated with cybersecurity, which drives rapidly evolving threat

---

[8] Modern Technology Solutions, "Simulation and Analysis Facility (SIMAF)," https://www.mtsi-va.com/modeling-simulation/, accessed August 10, 2021.

[9] U.S. Air Force, "Air Force Agency for Modeling and Simulation (AFAMS) Mission and Vision," https://www.mtsi-va.com/modeling-simulation, accessed August 10, 2021.

[10] U.S. Army Modeling and Simulation Office (AMSO), "AMSO's Mission," https://www.ms.army.mil, accessed August 10, 2021.

capabilities and corresponding changes in requirements for the system under development, as well as in the engineering and test infrastructures. Digital twins that can be subjected to repeated cyberattacks as the threats and their tactics, techniques, and procedures (TTPs) evolve will allow the cyber resiliency of the systems and mission to keep pace.

Use Uncertainty Quantification

The current DoD instruction on M&S verification, validation, and accreditation recognizes the importance of having and using a process to verify, validate, and accredit M&S tools (DoD, 2018b). Although such a process provides the necessary steps to ensure that a model meets the expectations of the service, the process is insufficient in that it does not take advantage of uncertainty quantification (UQ), a powerful tool for understanding the limitations of models.

A number of sources of uncertainty exist in modeling and simulation; a good discussion is found in Roy and Oberkampf (2011). Some uncertainty is the result of the modeling process itself and is introduced by modeling assumptions and numerical approximations employed in the simulation. Other sources of uncertainty result from the characteristics of the system itself, such as dimensional variations, variability due to manufacturing processes, wear, damage, and uncertainty in the system surroundings. UQ provides a framework to estimate the uncertainty of the result as well as the sources of those uncertainties. This information can be used to inform testing activities as to which tests may be most useful in driving down the uncertainty in the understanding of the system.

> **Recommendation 4-1. A Department of Defense joint program office should establish a shared, accessible, and secure modeling and simulation (M&S) and data ecosystem to drive development and testing across the life cycles of multiple supporting programs. M&S should be planned from early concept development to support the entire life cycle of the system, from requirements generation, through design development, integration and test, and sustainment. Uncertainty quantification should be employed to identify the primary sources of uncertainty in the understanding of the system being developed and to define an integrated testing and simulation activity to reduce those uncertainties to an acceptable level.**

The M&S ecosystem should:

1. Be shared within a DoD mission space for a set of critical mission threads;

2. Contain DoD validated and accredited scenarios, threat models, system models, and common metadata that defines the pedigree, applicability, and limitations;
3. Be accessible by concept developers, requirements developers, research and development programs, acquisition programs, and test facilities; and
4. Integrate across DoD services and industry partners so that industry models can be used in the ecosystem and DoD models can be used to support digital engineering by industry partners.

INCREASING THE USABILITY AND VALUE OF DATA

The role that digital technologies play in testing and evaluation is not limited to modeling and simulation. These technologies, for instance, make it possible to record, store, process, and analyze huge amounts of data from testing—data that can provide a much clearer and complete picture of the performance of a system or system of systems under test. Digital technologies also enable the rapid and secure communications and transfer of data. But taking advantage of these capabilities will require overcoming various challenges. Two of the biggest challenges will be handling massive amounts of data in a way that maximizes the value of that data and ensures the interoperability of data among the various segments of the OT&E establishment and across multiple programs in critical warfighting domains.

The Challenge of "Big Data" in Operational Testing

The growing power of digital storage—that is, the ability to hold increasing large amounts of data in increasingly small spaces and at increasingly low costs—combined with the increasing ability of computers to manipulate and analyze those data quickly and efficiently has created an era of "big data" in which previously unimaginable amounts of data are collected, stored, analyzed, and communicated. This is turn has revolutionized many fields that rely on large amounts of data, from artificial intelligence and machine learning (AI/ML) to autonomous systems, and it has the potential to have a similar positive effect on OT&E—if the data can be handled effectively.

In the January 2021 workshop, James Amato, the executive test director of the Army Test and Evaluation Command, observed that, as a result of the digital revolution, the military ranges are collecting more and more data from tests. "The amount of data that we push around and that we have to push between ranges, has grown exponentially," he told the committee. But the ranges are being overwhelmed by that data. "We don't have the [data] infrastructure today," he said. "We don't have the technology and solutions in place today to be able to do that at scale, at speed that will be required to link those."[11] Similarly, another speaker at the workshop, Arun Seraphin, a professional staff member of the Senate Armed Services Committee, identified the lack of an efficient data infrastructure as a major OT&E challenge. The ranges generate large volumes of test data, he said, but they do not manage those data well, and the main reason for that failure is that the ranges' data infrastructure is inadequate (NASEM, 2021, pp. 9–10).

As Conrad Grant, the chief engineer of the Johns Hopkins University Applied Physics Laboratory, explained at that workshop, there are a variety of data and measurement challenges in large-scale tests, such as those carried out across multiple domains or multiple ranges. "We need instrumentation, telemetry, data collection, data handling, and data analysis that will work at the scale of these large ranges we're talking about," he said, "and this is made difficult because of the desired volume of the data we're trying to collect from the system under test and the desire to make it available for analysis very quickly."[12] Speed is necessary, he noted, because evaluators often must analyze the data that have been collected from range tests on one day in order to determine which tests should be run the next day, but this speed is only possible if the large amounts of data collected from the tests can be quickly transmitted to the centers where the data analysis is done.

Also at that workshop, Joshua Marcuse, the head of strategy and innovation in the global public sector at Google, spoke about what will be necessary for the ranges to handle the large amounts of data generated by the tests. In particular, he argued that it is crucial to start planning for how those data will be handled early in the design phase of a project. However, he said, in his work with DoD, he has observed that program officers often design and build systems without a data strategy, with the result being that much of the most meaningful data—the data that can be used to inform operational testing—are not collected (NASEM, 2021, p. 9). "Thinking about the data requirements for a digital engineering approach to this has to begin at the beginning and not be a requirement that comes in at the end when the system is handed over the wall to someone that's meant to test it and then they realize what's missing," he said. This will require program officers to develop a new mindset, he said.[13]

---

[11] James Amato, presentation to Workshop on Assessing the Suitability of Department of Defense Ranges, January 29, 2021.

[12] Conrad Grant, presentation to Workshop on Assessing the Suitability of Department of Defense Ranges, January 28, 2021.

[13] Joshua Marcuse, presentation to Workshop on Assessing the Suitability of Department of Defense Ranges, January 29, 2021.

More generally, Marcuse told the committee at the workshop that military ranges lack the necessary digital resources to handle both the data-intensive and the computation-intensive aspects of OT&E. To properly carry out modeling and simulation of the sort required for testing and evaluation requires a tremendous amount of computing capacity—generally more than DoD has available (NASEM, 2021, p. 9). He continued on to observe that some military ranges hardly seem to have entered the digital age at all, and he spoke about a time he was at an Army testing facility where people "were complaining to us enormously because they had a difficult time keeping track of all the paper copies of the testing results that they needed to get from the range that they were supposed to inspect."[14]

Data Communication Issues

One particular challenge related to the vast amounts of data that will be generated by the tests of the future is simply moving those data from one place to another among the relevant platforms, range assets, and participating test ranges. This requires a highly connected, high-capacity, highly secure communications system that is far beyond anything that exists in the nation's system of military ranges today, and the requirements will only increase as time goes on, testing becomes even more sophisticated and data-intense, and the need for rapid and dependable communication of data grows. Addressing these challenges will require improvements in both hardware and software, and the particular issues related to data communication range from simple infrastructure needs (establishing connectivity, increasing data rates, expanding communications frequencies and data formats) to more complex issues such as a lack of standardized data formats and conflicting information security approval authorities.

A recent effort supported by DoD to enable the fast transfer of large volumes of data is the Defense Research and Engineering Network (DREN).[15] DREN is a fiber optic network connecting supercomputing centers for scientific research as well as test and evaluation missions. A secret version of DREN, the SDREN, provides a network for transferring secret level data. While this could be a promising effort for improving intra-range connectivity for complex test events, it is unclear if DREN or

---

[14] Ibid.

[15] Defense Research Engineering Network (DREN)/Secret Defense Research Engineering Network (SDREN), "Network Capabilities and Technical Overview," https://www.hpc.mil/program-areas/networking-overview/dren-sdren, accessed August 18, 2021.

SDREN can accommodate data transfers at multiple classification levels and if they resolve issues with data interoperability.

Data Security Issues

DoD has unique needs for data security. Test range data is commonly a mix of security classification levels and needs to support sharing via a multi-level security mode of operation and not simply be defaulted to a "system high" mode of operation. The data generated at the test range may also be proprietary. Properly facilitating the sharing of required information is both a technical issue and a computer security/bureaucratic approval issue. The sharing of data, models, and other digital assets among ranges and among services is going to become increasingly important in coming years, but such sharing leads to a number of security issues. As Ed Greer, former Deputy Assistant Secretary of Defense for Developmental Test and Evaluation, said at the public workshop, it is difficult to share data securely among various entities because of the lack of a "robust common IT [information technology] infrastructure that can support multi-level security and the switching of classification levels quickly" (NASEM, 2021, p. 10).

The committee site visits revealed specific examples of security issues related to testing and sharing of information. For example, the Air Force Capability and Encroachment Assessment Detail at the Eglin Test and Training Complex needs T&E infrastructure upgrades to support next-generation testing. The range cannot support the multi-level classification needs for the T&E environment. Furthermore, net-centric warfare requires realistic test environments for systems-of-systems interoperability (Figure 3-36 in DoD, 2018a), which will further exacerbate these security issues.

In speaking with personnel at the White Sands Missile Range (WSMR), the committee learned that the Test Resource Management Center (TRMC), the Survivability/Lethality Analysis Directorate at the Army Research Laboratory, and other DoD resources provide assistance to WSMR and other test facilities to help secure their existing cyber systems as well as to assist in the creation of high-fidelity, mission-representative cyberspace environments for testing and evaluation. In order for WSMR to maintain up-to-the-minute awareness of cybersecurity and cyber T&E advances, WSMR leadership must bring together a multi-directorate group to create a roadmap forward (WSMR, 2016).

Data Interoperability and Security Challenges to Sharing Data

In addition to the basic challenge of moving huge amounts of data from place to place, as described above, military ranges face two other challenges related to moving data—and other digital resources, such as models—from range to range and system to system quickly and securely. Specifically, this sort of sharing is limited by two basic issues: limitations in data interoperability and difficulties in ensuring security when transferring and sharing data.

Data Interoperability

The current lack of data interoperability among ranges and systems has its roots in a variety of factors. To begin with, legacy systems, which generally have been developed with unique data definitions, pose a major challenge to interoperability. Individuals in different places and at different times made choices about their data that were tailored to fit their own particular requirements without much, if any, concern about whether those data could be combined or compared with data generated by others making decisions about their data based on very different considerations. The result is that the ranges have a mishmash of different data systems with varying data definitions and formats. Even today, when the value of data interoperability is more widely recognized than in the past, the designers of individual systems will often make locally optimal decisions about data definitions and formats. The result is that the various data systems operating on military ranges have limited data interoperability. This can be overcome—with some effort—when the goal is to improve the data interoperability between two systems or among a small number, but the task becomes more and more complicated as more systems are involved, and the issue is most apparent regarding the data interoperability of complex systems of systems.

DoD has been aware of this issue for quite some time, and, indeed, in the 1990s the department launched two major efforts to address application interoperability with the goal of preserving meaning and being mutually interpretable (NRC, 1999). The first of these efforts was the Enterprise Data Model Initiative, which sets forth a DoD process through which standard data definitions in functional areas (e.g., command, control, communications, computers, and intelligence [C4I]; logistics; and health care) are developed and then subjected to a cross-functional review process prior to being adopted as DoD standards (DoD, 1994). The second was the Shared Data Environment (SHADE) Program, which enables different C4I systems to share data segments and to use standardized access methods using middleware for translating data elements from one system for another (DISA, 1996).

Personnel from the Nevada Test and Training Range spoke with the committee about how there are instrumentation challenges in providing fourth- and fifth-generation aircraft with encrypted capability.

This requires costly instrumentation infrastructure on the aircraft and in ground support. In the *2025 Air Test and Training Range Enhancement Plan* (USAF, 2014) it was noted that the Common Range Integrated Instrumentation System (CRIIS) project will provide most major range test and facility bases with the capability to collect highly accurate time, space, position information, and selected aircraft data bus information needed for advanced weapon systems testing. The enhancements provided by CRIIS are expected to enable interoperability across the major test ranges and support future F-35 testing (DoD, 2018a).

While the elements of DoD's strategy for achieving interoperability are positive, they are not being fully executed. A 1999 study from the National Research Council found that both the formulation and the implementation of this strategy had gaps and shortfalls (Finding I-1 from NRC, 1999). And according to what committee members heard from staff at TRMC, data interoperability continues to be an issue more than two decades later. Box 4.1 provides a sample of data challenges voiced at the workshop (NASEM, 2021).

To ensure the usability and value of the data collected on the nation's military ranges, the committee makes the following recommendations:

BOX 4.1
Summary of Data Challenges for Test Ranges

Listed below is a sample of data challenges voiced by members of the test community to the committee at its public workshop:

- Ranges need to work with customers to be ready for large exercises and ensure that all parties are adhering to data standards.
- It will be necessary to deal with large volumes of data gathered from multiple sources and make it available for analysis quickly.
- Many ranges lack sufficient bandwidth and clear protocols for the real-time transfer of test data generated at various classification levels.
- Need to standardize data exchanges and standardize instrumentation. This would enable ranges and program managers to spend less time and effort to correlate data from different ranges.
- Programs that build systems do not have data strategies. Programs don't have the data you need or want, data storage computation, or capacity.
- Currently there are large amounts of test data that are being generated but not used.
- A barrier to data sharing and analysis is security challenges among the test ranges. Security challenges include the program and range investments in system and network certification and accreditation.

SOURCE: NASEM (2021).

Recommendation 4-2: A Department of Defense joint program office should adopt and promulgate modern approaches for standardization, architectural design, and security efforts to address data interoperability, sharing, and transmission challenges posed by the complexity of next-generation systems. The joint mission office should determine how to develop and maintain a protected data analysis tool and model repository for testing, increase the interconnectivity of test ranges, and ensure the development of data protocols for the real-time transfer of data at multiple classification levels.

Few ranges have sufficient bandwidth and clear protocols for the real-time transfer of test data generated at various classification levels. For data that are not prioritized for real-time transfer, the transfer can take weeks to reach appropriate analysts, potentially resulting in significant scheduling delays. A pragmatic phased adoption approach will need to take account of the maturity of data tools and processes, and that implementation will require both up-front investment and concerted effort.

Software Is a Challenge in Operational Testing

With the digital revolution, software has become an increasingly important part of military weapons and systems, to the point that today's systems, from the F-35 to artillery, are almost completely dependent on the proper functioning of their software. This means that the testing and evaluation of military systems includes a major software testing component. However, what the committee found from its study is that the military's testing ranges have not kept up with software capabilities.

For instance, in the workshop sponsored by the committee as part of this study, Marcuse said that a fundamental challenge facing DoD is that, despite the digital revolution, testing remains optimized for hardware. The implications of that revolution, Marcuse said, have not permeated DoD's rules, processes, institutions, or its personnel (NASEM, 2021). Going forward, military testing and evaluation should focus more on the digital elements of systems. The committee heard similar testimony from Raymond O'Toole, the acting director of OT&E at DoD, who said that "dramatically increasing and improving the test and evaluation of software-intensive systems" should be one of DoD's priorities in OT&E (NASEM, 2021, p. 4). And Seraphin told the committee, "We have a real concern over the department's ability to test software, both on the workforce side and on the infrastructure side."[16] Seraphin pointed to

---

[16] Arun Seraphin, presentation to Workshop on Assessing the Suitability of Department of Defense Ranges, January 29, 2021.

a number of specific software areas as presenting challenges in testing and evaluation, including software for emerging AI systems, software for command-and-control systems, and software for business systems.

Testing Artificial Intelligence and Autonomous Systems

The greatest software challenge for OT&E—and for T&E in general—is likely to be in the area of AI software and AI-based autonomous systems, as the committee heard from a number of sources. AI and autonomous systems are expected to play a major role in the nation's defense in coming decades (Ray et al., 2020), but, to date, relatively little has been done to prepare for the testing of such systems. At the workshop, for instance, Jane Pinelis, the chief of testing, evaluation, and assessment at DoD's Joint Artificial Intelligence Center (JAIC), said that the military's testing and evaluation capabilities "have not been keeping pace with the speed of AI technology development" (NASEM, 2021). And in site visits to various ranges, committee members heard on multiple occasions that the ranges are completely unprepared to test systems running AI, including autonomous systems.

There are multiple reasons why the testing of AI and autonomous systems is challenging for the ranges. This is a technological area in which rapid progress is being made, which means that it is difficult to anticipate and prepare for the sorts of systems that might employ AI and to predict what the capabilities of those systems might be. But, to a degree, this is true about any technology in which rapid advances are being made. However, because of the nature of AI and autonomous systems, they pose testing challenges that are unlike any other.

For example, Devin Cate, the director of test and evaluation for the U.S. Air Force, told the committee during the workshop that because AI and autonomous systems are learning systems, they inevitably change and evolve throughout testing, which makes it difficult to characterize their performance in a repeatable manner. Overcoming this issue, he suggested, will require the testing enterprise to work closely with the system developers so that the AI-enabled and autonomous systems are designed from the start with tests in mind; in particular, he suggested, it would be useful to design these systems to collect all the data that will be needed to characterize and judge their performance (NASEM, 2021).

Another testing challenge will be simply setting performance goals for these systems since it is difficult to make a connection between specific performance parameters of the systems and the outcome of operational or mission tests. Things get even more complicated when the testing involves humans teaming with AI or autonomous systems. It will be critical to do such integrated tests in order to evaluate how the systems will perform in actual missions, but at present there is no well-established approach for carrying out tests of such combinations.

Perhaps the most challenging aspect to testing AI and autonomous systems will be determining how to detect and evaluate emergent behaviors—actions that the systems take that have not been programmed into them but rather that appear as the result of complex interactions among a system's various components or because of machine learning. A non-military example would be the selection of a chess move by an AI chess system—the machine chooses its moves through its own study of chess, and the machine's creators have no idea what a move will be until it has been made. The performance of a chess-playing computer can be judged by, for instance, having it play multiple games against human grandmasters (or against other chess-playing computers). It is not clear, however, how to judge the emergent behavior that will appear in AI-enabled military systems. As Pinelis told the workshop, "We need methods for defining, diagnosing, and understanding emergent behavior as well as human training so that the operator can identify emerging behavior as it occurs and do things about it if it is undesirable."[17]

A related issue will be how to judge a particular performance—to decide what is "passing"—when an AI-driven system is being evaluated. This was mentioned at the workshop by Marc Bernstein, the chief scientist under the Assistant Secretary of the Air Force for Acquisition, Technology, and Logistics. As an example, he pointed to the Advanced Battle Management System (ABMS) now being developed by the Air Force. Evaluating the ABMS properly will require testing it in complex environments where there is no single "correct" action but rather a collection of options, each with its own advantages and disadvantages, so that the "best" choice is a judgment call. How, he asked, do you set up your operational testing and evaluation in such an ambiguous, gray environment? Furthermore, given that AI systems do their own "thinking" and do not simply behave in ways that have been programmed into them, it is quite possible that the AI-enabled system will come up with an optimal solution that is different from what its evaluators believe is best—and perhaps it would even come up with a solution that its evaluators had never thought of—and in these cases it can be difficult, if not impossible, to judge the system's performance accurately (NASEM, 2021, p. 3).

Yet another issue was pointed out at the workshop by Grant. In testing weapon systems on autonomous vehicles where the weapons may be under the control of AI, how can the safety of others on the ranges

---

[17] Jane Pinelis, presentation to Workshop on Assessing the Suitability of Department of Defense Ranges, January 29, 2021.

be assured, given that the AI's decisions are not generally predictable? (NASEM, 2021).

Given all of these considerations and the fact that AI-enabled systems under test can have some very severe consequences, Pinelis told the workshop that it is crucial that DoD "push the test and evaluation for AI-enabled systems to where it needs to be with respect to science, data, knowledge, skills, workforce, and infrastructure" (NASEM, 2021, p. 3). At the same workshop, Missy Cummings, professor in the Department of Electrical and Computer Engineering at Duke University, offered a sobering warning about the difficulty of modeling autonomous systems. "Simulation can maybe help you do some baby testing early in the phases of autonomous systems, but it simply cannot represent the uncertainty of the real world" (NASEM, 2021, p. 9). In a review of the literature and site visit discussions, the committee found that the ranges are not adequately prepared for the testing and evaluation of AI and autonomous systems.

> **Finding 4-1:** DoD test ranges are unprepared for the operational testing and evaluation of the increasing integration of AI and autonomous systems in military systems.

In an effort to develop a collaboration platform to support autonomy and AI projects and programs for DoD, the Office of the Under Secretary of Defense for Research & Engineering established the Assured Development and Operation of Autonomous Systems (ADAS) Project, which is overseen by TRMC.[18] ADAS was initiated in 2020 to solicit proposals for making data, DevSecOps, software, and infrastructure resources accessible for collaborative settings to support autonomy and AI projects. To enable seamless collaboration across the services and domains for AI and autonomous systems testing, the committee makes the following recommendation:

> **Recommendation 4-3: The Test Resource Management Center should continue monitoring and supporting the Assured Development and Operation of Autonomous Systems Project, and prioritize efforts to develop a common set of standards, measurement approaches, and operational scenarios from which to evaluate the performance of artificial intelligence (AI) and autonomous systems, while recognizing that testing approaches may differ between AI and autonomous systems.**

---

[18] Arcnet Consortium Press Release, June 5, 2020, https://www.arcnetconsortium.com/trmc-coeus-white-paper-request/.

It is critical that program managers, TRMC, and DOT&E recognize that next-generation systems that continuously evolve as a result of changing data, AI integration, and similar technological advancements will require new methods for testing. For example, ABMS processes large volumes of data to inform decision making on the joint domain battlefield. Alternatively, autonomous systems require more work in the integration of human-machine teams. Given that changes in data will result in different outputs, testing the evolving ABMS may require continuous operational testing exercises year after year to ensure its suitability and survivability. Further research is necessary to advance testing technologies and strategies to test the integration of AI and autonomous systems.

REFERENCES

Biesinger, F., B. Krab, and M. Weyrich. 2019. "A Survey on the Necessity for a Digital Twin of Production in the Automotive Industry." 23rd International Conference on Mechatronics Technology (ICMT). doi: 10.1109/ICMECT.2019.8932144.

DAU (Defense Acquisition University). n.d. *Modeling and Simulation (M&S) and Distributed Testing*. Fort Belvoir, VA. https://myclass.dau.edu/bbcswebdav/institution/Courses/Deployed/TST/TST303/Student_Materials/Student%20Lessons%20%28PDF%29/L05S-Model%20%26%20Sim/L05-M%26S. Accessed May 5, 2021.

DISA (Defense Information System Agency). 1996. "Defense Information Infrastructure (DII) Shared Data Environment (SHADE) Capstone Document." Fort Meade, MD.

DoD (Department of Defense). 1994. *The DoD Enterprise Model. Volume 1: Strategic Activity and Data Models*. Office of the Secretary of Defense. January. https://www.hsdl.org/?view&did=469009.

DoD. 2018a. *Report to Congress on Sustainable Ranges*. Washington, DC.

DoD. 2018b. "Department of Defense Instruction Number 5000.61." https://www.esd.whs.mil/Portals/54/Documents/DD/issuances/dodi/500061p.pdf.

NASEM (National Academies of Sciences, Engineering, and Medicine). 2021. *Key Challenges for Effective Testing and Evaluation Across Department of Defense Ranges: Proceedings of a Workshop—In Brief*. Washington, DC: The National Academies Press.

NRC (National Research Council). 1999. *Realizing the Potential of C4I: Fundamental Challenges*. Washington, DC: The National Academies Press.

Ray, B.D., J.F. Forgey, and B.N. Mathias. 2020. "Harnessing Artificial Intelligence and Autonomous Systems Across the Seven Joint Functions." *Joint Force Quarterly* 96:115–128.

Reis, V., R. Hanrahan, and K. Levedahl. 2016. "The Big Science of Stockpile Stewardship." *Physics Today* 69(8):46. https://physicstoday.scitation.org/doi/10.1063/PT.3.3268.

Roy, C.J., and W.L. Overkampf. 2011. "A Comprehensive Framework for Verification, Validation, and Uncertainty Quantification in Scientific Computing." *Computer Methods in Applied Mechanics and Engineering* 200(25-28):2131–2144. https://www.sciencedirect.com/science/article/abs/pii/S0045782511001290.

Siegfried, R. 2021. "Special Issue: Modeling and Simulation as a Service." *Journal of Defense Modeling and Simulation: Applications, Methodology, Technology* 18(1).

USAF (U.S. Air Force). 2014. *2025 Air Test and Training Range Enhancement Plan*. Report to Congressional Committees. http://www.nttrleis.com/documents/review/2025%20Air%20Test%20and%20Training%20Range%20Enhancement%20Plan_Jan2014.pdf.

USFDA (U.S. Food and Drug Administration). 2021. "Model Informed Drug Development Pilot Program." https://www.fda.gov/drugs/development-resources/model-informed-drug-development-pilot-program. Accessed August 5, 2021.

WSMR (White Sands Missile Range). 2016. *White Sands Missile Range 2046 Strategic Plan*. https://www.wsmr.army.mil.

5

Speed-to-Field: Restructuring the Requirements and Resources Processes for DoD Test Ranges

Delivering new systems into the field as quickly as possible should be one of the main goals of the Department of Defense (DoD) test and evaluation (T&E) system, but that speed should not come at the cost of losing the rigor of that T&E. Satisfying both of these conditions—enabling speed-to-field while maintaining the rigor of DoD's operational test and evaluation in today's world of highly complex technologies—is becoming increasingly difficult for today's DoD ranges. Indeed, a large portion of DoD's T&E ranges were developed in the 1950s and suffer from outdated equipment that is expensive to maintain and increasingly inadequate to meet the testing demands of the future, such as hypersonic weapons and multi-domain operations (MDOs). Many of DoD's ranges will require substantial investments in modernization just to meet current operational testing needs and much more to prepare for the future.

This chapter summarizes the challenges with current range infrastructure investments processes and lays a foundation for Congress and DoD work together to ensure appropriate funding to modernize and recapitalize DoD ranges in order to meet the needs of the modern battlefield and the intricacies of modern weapon systems.

The emphasis of this chapter is how DoD can better determine current and future testing needs, evaluate the existing range capabilities, identify range facility shortfalls, develop strategies to fund both current operations and long-term capital investments, and improve speed-to-field. In order to provide adequate funding, requirements for test and evaluation of weapons and systems must be accurately defined early in the program acquisition cycle. Testing discussions often occur late in program development, meaning that testing gaps are identified late, and the ranges may not be equipped to conduct appropriate testing. Understanding how test ranges currently fund modernization efforts and prepare to meet testing requirements is critical for developing new strategies to improve the responsiveness, effectiveness, and flexibility of the test enterprise.

PROGRAM REQUIREMENTS DRIVE RANGE FUNDING INVESTMENTS

Test range operating and maintenance costs are funded by their owning service, but some execution costs are reimbursed by the test customers. Each program has a set of T&E requirements that are typically established in the acquisition cycle of the program. The program manager and testers must ensure that test resource requirements are identified early in the acquisition cycle, that they are documented in the initial test and evaluation master plan (TEMP), and that modifications and refinements are reported in the TEMP updates. The services make test resource decisions based on what they predict they will need in 3 to 5 years according to the requirements documents in the TEMPs. Once testing needs are prioritized and established, they are difficult to modify. It is critical to note there is no stable funding process where program funds are put aside for future OT&E needs.

Funding for recapitalization and modernization of the ranges is driven by the testing requirements set in the acquisition process of the program. If a service Other Transaction Authority (OTA) or the Director of Operational Test and Evaluation (DOT&E) sets an operational testing requirement calling for capabilities not yet available at a test range, the range works with the program manager, the range sponsor, or the Test Resource Management Center (TRMC) to obtain the necessary funding and develop the capability to meet the testing requirement.

There are a variety of drawbacks for operational testing from the current piecemeal requirements process. A lack of defined and achievable requirements at initial approval of the program test strategy results in unreliable cost and schedule estimates. The committee also observed that program funding to support the ranges results in a large number of individual capability projects, but there are few resources to develop infrastructure to connect and integrate these capabilities. Additionally, if program requirements are the driving force for prioritizing range investments, then the ranges are not preparing for the next-generation capabilities that may be needed in the next 10 to 15 years.

Finding 5-1: Program test requirements frequently drive funding decisions for range modernization, so recapitalization and modernization for broader testing use is not incentivized in the current funding structure.

COLORS OF MONEY FOR RANGE MODERNIZATION AND MAINTENANCE

Currently, there is a complex funding stream to operate tests, address deteriorating test assets, and modernize test ranges. During the site visits, the committee had discussions with range personnel that were aimed at developing a better understanding of range funding for modernization and sustainment. The funding profiles of the test ranges are not identical, but in general they rely primarily on reimbursables from their customers or indirect costs through appropriations. Figure 5.1 illustrates the various sources of funding for the test ranges, which are further decribed throughout this chapter. However, consistent and aggregated data appears to be

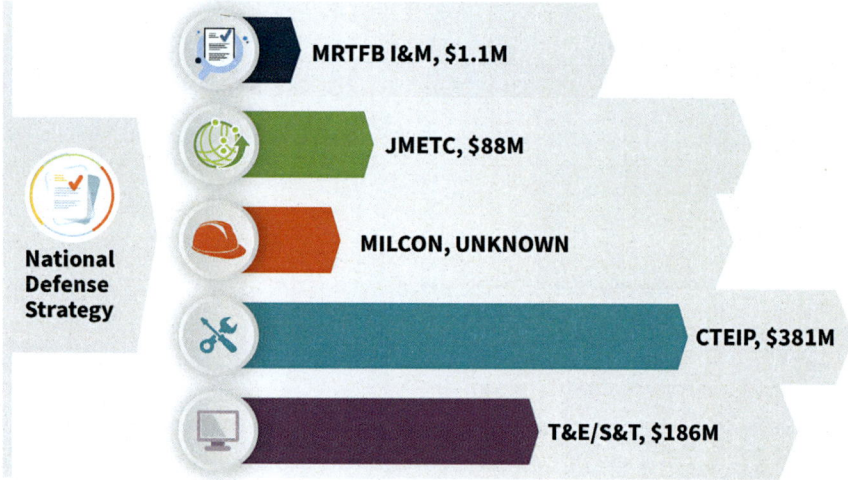

FIGURE 5.1 Department of Defense (DoD) institutional funding sources for test ranges, including Major Range and Test Facility Base Investments and Modernization (MRTFB I&M) funding, the Joint Mission Environment Test Capability (JMETC) program, military construction (MILCON), the Central Test and Evaluation Investment Program (CTEIP), and the Test and Evaluation/Science and Technology (T&E/S&T) program. SOURCE: Budget amounts for JMETC, CTEIP, and T&E/S&T from the Office of the Secretary of Defense, "Budget Estimates for Fiscal Year 2021 (Department of Defense Fiscal Year 2022 Budget Estimates," https://comptroller.defense.gov/Portals/45/Documents/defbudget/fy2022/budget_justification/pdfs/03_RDT_and_E/RDTE_Vol3_OSD_RDTE_PB22_Justification_Book.pdf, accessed September 9, 2021. Budget amount for MRTFB I&M provided by TRMC on request.

unavailable on how funding for range modernization, operation, and maintenance is allocated, what the overall requirements are, as well as the resulting capability gaps and deferred maintenance levels.

The 23 major range and test facility bases (MRTFBs) secure a significant portion of their approximately $4 billion funding through indirect costs paid via appropriations, but they are restricted by law from recovering indirect costs from the majority of their customers. However, non-MRTFBs can secure indirect costs from customers, which can be a substantive portion of their funding profile.

Resources specific to a particular test must often be developed and funded from program managers' research, development, test, and evaluation (RDT&E) budgets (CRS, 2020a,b). Program managers and testers must ensure that test resource requirements are identified early in the acquisition cycle and that they are documented in the initial TEMP. Requests for test resources are also outlined in the TEMP.

Range Funding from Service Programs

Each of the services has an understanding of its own test ranges and capabilities and is tasked with managing and operating its designated MRTFB activities. DoD Defense Directive 3200.11 and DoD Instruction 3200.18 state that the TRMC reviews and certifies the proposed T&E budgets.[1] The services must also coordinate any proposed changes to T&E capabilities and infrastructure with TRMC before making those changes. The funding processes from the services to support tests at the ranges differ, but brief descriptions of those processes are provided to illustrate their complexity.

The Army Test and Evaluation Command (ATEC) oversees eight testing locations and provides direct support to Army Futures Command for independent operational testing and evaluation. Funding for Army operational testing is through the program managers and directed to ATEC for the control of the funds. Weapon system program managers use RDT&E funds to reimburse supporting commands for costs directly related to their tests.

The Air Force Operational Test and Evaluation Center (AFOTEC) is the Air Force's agency responsible for operational testing over five detachments and three operating locations for programs on DOT&E oversight. AFOTEC has direct input on OT&E funds for all Air Force programs. Costs associated with the initial operational test and evaluation (IOT&E)

---

[1] Department of Defense (DoD), "Major Range and Test Facility Base (MRTFB)," DoD Directive 3200.11, December 27, 2007, https://www.esd.whs.mil/Portals/54/Documents/DD/issuances/dodd/320011p.pdf?ver=2018-10-24-083959-987.

are RDT&E-funded, and the costs of OT&E are funded with operations and maintenance (O&M) funds.

The Commander, Operational Test and Evaluation Force (COMOPTEVFOR) commands the Navy's independent OT&E activity and reports directly to the Chief of Naval Operations (CNO). CNO funds the development of generic test resources for use in OT&E, but the program manager uses the program's RDT&E funds to support the support the execution of the test program.

Investment Programs to Support Range Modernization

For priority areas listed in the National Defense Strategy, the services and the test ranges have additional resources outside the program funds for building test capabilities and infrastructure. TRMC administers approximately $500 million in investments annually to address shortfalls in T&E capabilities. Those investments are spread out across three investment programs, briefly described below.

The mission of the Central Test and Evaluation Investment Program (CTEIP) is to develop or improve major test capabilities that have multi-service utility. TRMC administers CTEIP through a corporate investment approach to combine Service, Defense, and other government agencies T&E needs, to maximize opportunities for joint efforts, and to avoid unwarranted duplication of test capabilities. CTEIP focuses investments on projects that will have high productivity returns on investment. Projects under the CTEIP historically supported two basic tasks: investments to improve the test capabilities base (Joint Improvement and Modernization [JIM] projects) and the development of near-term solutions to test capability shortfalls in support of ongoing operational test programs (Resource Enhancement Project [REP]). The services typically use a competitive process to determine how to use CTEIP funding to pay for test range investments. The test ranges themselves track their investment needs through their internal strategic planning processes and the services guide the funding decisions.

Another TRMC-administered investment program is the Test and Evaluation/Science and Technology (T&E/S&T) Program. T&E/S&T is approximately a $100 million investment program established in 2002 to exploit new technologies and expedite their transition from the laboratory to the T&E community. Currently, test technology areas include cyberspace, directed energy, electronic warfare, high-speed systems, net-centric systems, unmanned and autonomous systems, advanced instrumentation systems technology, and spectrum efficient technology.

The Joint Mission Environment Test Capability (JMETC) program prioritizes interoperability by providing funds for robust distributed infrastructure (network, enterprise resources, integration software, tools, reuse repository) and technical expertise to integrate live, virtual, and constructive (LVC) systems for test and evaluation in joint systems-of-systems and cyber environments.

There are a few additional resources available to test ranges that can be applied to modernization efforts. Test ranges can obtain funds from the military construction (MILCON) program,[2,3] MRTFB Institutional (direct appropriations), the Spectrum Relocation Fund, the Readiness and Environmental Protection Integration (REPI) program, and MRTFB Investment & Modernization (I&M) funding. Furthermore, RDT&E and O&M funds can be used for unspecified minor military construction projects to support test and evaluation activities, for projects costing not more than $6 million (Title 10 U.S. Code, Section 2805 – Unspecified minor construction). This amount was raised in 2017 and alleviates total reliance on the slow and uncertain process for obtaining MILCON funding. However, during the committee site visit to the Atlantic Test Range (ATR), personnel indicated that the $6 million limit forces the construction of low-cost or temporary structures and does not address significant facility refurbishment costs that generally exceed the limit. ATR personnel also said that additional range modernization would be possible with their existing level of resources but that the severe limitations on mixing funding streams cannot accommodate shifting priorities or emerging test needs.

Maintenance and Repair of Test Ranges

Many test facilities with aging infrastructure still have high usage rates. The Strategic Plan for Department of Defense T&E Resources from March 2013 noted,

> Due to age and outmoded technology, many test facilities are increasingly difficult to sustain and/or maintain. Obsolescence and deterioration contribute significantly to increased levels of maintenance, reductions in reliability, and an overall increase in operating costs. Services are under pressure to keep existing ground test facilities viable and relevant to meet immediate and forecasted needs. Across all services, there has been a downward trend in T&E military construction (MILCON) appropriations to address ongoing maintenance, sustainment, and modernization needs of our T&E facilities. Further analysis is required (e.g., recapitalization rate) to provide a comprehensive assessment of MRTFB-only MILCON needs and investments.[4]

---

[2] DoD Directive (DoDD) 4270.5.

[3] U.S. Code Title 10 - Chapter 169: Military Construction and Military Family Housing.

[4] U.S. House of Representatives, House Report 114-102 to accompany House Report 1735, p. 356, 2015, National Defense Authorization Act for Fiscal Year 2016, https://www.congress.gov/114/crpt/hrpt102/CRPT-114hrpt102.pdf.

Funding to sustain and support aging infrastructure of legacy capabilities is strained based on new capability development and the sustainment costs of any new effort. This results in deferred maintenance, which can result in increased costs for test operations as well as higher costs to replace the capability compared with costs associated with earlier mitigation. Personnel from Eglin and Edwards Air Force bases highlighted concerns regarding funding shortfalls for maintaining legacy systems. Personnel from Naval Air Station Point Mugu spoke of how nearly one-third of their I&M budget goes directly toward maintenance on existing infrastructure, and personnel from Point Mugu pointed out that it is ultimately the customers who bear the financial brunt of aging infrastructure costs. The Air Force Test Center reported in 2021 that current funding is insufficient for maintaining critical test facilities, including wind tunnels and anechoic chambers (Department of the Air Force, 2021). The report states that the insufficient funding for these facilities results in a high risk for failure and a reduction in the ranges' capabilities and capacities.

Finding 5-2: There are inadequate funds for the maintenance and sustainment of DoD test range infrastructure and capabilities, and costs due to deferred maintenance continue increasing.

According to a Government Accountability Office report, DoD reported approximately $100 billion in deferred maintenance and repairs across all DoD facilities between FY2009 and FY2014, and further determined that over those years DoD spent only 79 percent on average of the estimated facilities maintenance requirements (GAO, 2016). While this report covers more than DoD ranges only, the fact that GAO's site visits included Eglin Air Force Base and Aberdeen Proving Ground suggest that ranges may also be underspending on deferred maintenance. As a result, prior budgetary allocations for facility maintenance and sustainment are unlikely to address current and projected deferred maintenance costs. If the ranges' capabilities and infrastructure are not adequately maintained, they cannot achieve optimal performance and may have a reduced service life, which directly affects operational test schedules, program and range budgets, and program mission.

STRATEGIES TO IMPROVE TEST RANGE MODERNIZATION

If test and evaluation processes are initiated in the formative stages of a program, the test ranges can provide feedback on the test requirements based on available range capabilities and resources. Should the program tests require new capabilities or infrastructure repair, this strategy will maximize the time available to the ranges to prioritize investments in modernization and maintenance. Additionally, these discussions are key for determining how M&S can be used for test as well as the types and volume of data required and generated during test. These elements are key for accurately determining investment needs for test instrumentation, infrastructure, personnel, and timing. Bringing OT&E into the program acquisition cycle early will support the:

1. Identification of appropriate test requirements.
2. Identification of any range shortfalls in testing capabilities.
3. Establishment of funding streams to ensure the ranges will be ready to do appropriate testing when the system is ready to be tested.
4. Examination and facilitation of synchronization between operational and developmental testing requirements early in the acquisition process.

Given the multi-domain dimensions of the battlespace and the emergence of connected and concurrent kill chains, many programs will also need to be tested in an integrated environment, where the program interacts with other systems and across domains. Preparing a program for multi-domain testing requires discussions across the services. The Joint Requirements Oversight Council (JROC), which charters and oversees efforts to develop joint operational and integrating concepts for joint missions during joint concept development, provides a unique opportunity for the services to examine validated test infrastructure requirements for connected concurrent kill chains and MDOs. Currently, the role of DOT&E is to serve as an advisor to JROC and its subordinate boards and coordinate with the Functional Capabilities Boards in its endorsement of Joint Capabilities Integration and Development Systems (JCIDS) (CJCSI, 2018). JCIDS provides the baseline for documentation, review, and validation of capability requirements across the Department. The committee supports this collaboration and seeks to add clarity to the goals of the collaboration through the following recommendation:

Recommendation 5-1: The Joint Requirements Oversight Council (JROC) should consult regularly with the Director of Operational Test and Evaluation (who is an advisor to the JROC) about the test requirements for systems considered by the JROC. This consultation should include an evaluation of current testing capabilities, facilities shortfalls, and plans to address these shortfalls.

Given DoD's annual programming and budgetary cycles, an annual report on the evaluations and the JROC outcomes could be timed to align with the annual acquisition Program Review. The Program Review is a time-based event where key stakeholders on an acquisition program gather to discuss the progress of their program.

For several next-generation DoD technologies, test capabilities still need to be developed. As an example, replicating environments for hypersonic systems is challenging, given their long flight distances and unique physical situations, including extreme temperatures in flight and impact on surrounding air flow. Designing facilities to test hypersonic systems can be costly and as challenging as designing the system itself.

The Office of the Undersecretary of Defense publishes a baseline standard for cumulative obligation and expenditure rates, and included in that publication are standards for RDT&E. The main goal of the practice is to ensure that DoD spends the funds appropriated by Congress in a timely manner. However, the projected expenditure rates are progressively failing to meet the execution benchmarks for many accounts (Conley et al., 2014). These projections are important because the timing of the programs can negatively affect test range investments, making it challenging to have enough funding to develop appropriate capabilities. A recent analysis on the reliability and accuracy of projected expenditure rates concluded that services do not appear to be planning or expecting to meet benchmarks from the onset of the program's budget process (Daniels & Harrison, 2020).

Test infrastructure to integrate and validate new technologies requires customization and long lead times on infrastructure preparation, which is not executable within projected benchmarks, especially for expenditure rates. Test investment programs historically achieve expenditure benchmarks in their third year, and applying Office of the Secretary of Defense (OSD) expenditure benchmarks to the beginning of a test modernization effort can place critical test technologies at risk of not getting started or maintaining funding as well as limiting range modernization to test advanced technologies. This limits DoD in its ability to initiate and complete required range modernization.

Recommendation 5-2: The Office of the Secretary of Defense should either allow an exemption or set shallower expenditure benchmarks for the first 2 years of test modernization programs. This will reflect realistic expense curves for the technologies and projects needed to test next-generation programs and complex integration.

In addition, the ability to perform minor military construction has proven invaluable in enabling OT&E in spite of the uncertainties in the MILCON process, but the current limit of $6 million over-constrains the military services and DoD, given the typical costs of even modest test infrastructure construction and facility refurbishment. Congress has previously authorized minor MILCON via RDT&E or sustainment funding in Title 10 U.S. Code, Section 2805, but the current approval process and limitations on project size prevent the use of an effective tool for addressing many range shortcomings.

> **Conclusion 5-1:** New mechanisms and funding limits for applying minor military construction are necessary for responsive test and evaluation activities.

Once testing requirements are better and earlier defined, shortfalls are identified, and funding requirements are calculated, there must be better and simpler ways not only to fund current test operations, but also to ensure needed recapitalization and modernization of DoD's ranges. Testing needs will also likely shift during the program engineering and development process as well as from the increasing sophistication of testing technologies. For example, a program may identify a data analysis platform or virtual environment early in the acquisition process, but advances in those technologies could affect testing costs. Additionally, as acquisition cycles become shorter, test ranges need to access resources for modernizing their capabilities quickly. This likelihood was raised during the Atlantic Test Range site visit and illustrates the lack of program funding agility to accommodate shifting OT&E needs. Therefore, ranges need flexibility to move investments to accommodate testing changes.

> **Finding 5-3:** Resources for test ranges to modernize their capabilities quickly are currently inadequate as acquisition cycles are becoming shorter and testing needs shift over the course of project development.

> **Conclusion 5-2:** There exists a need for the Department of Defense to pilot new processes and authorities for funding ranges and infrastructure to make them simpler, more responsive, and more effective.

To this end, DoD could conduct a pilot study, using one of the emerging technologies identified in the DoD National Strategic Plan, and determine the adequacy of the current ranges to provide needed test and evaluation; identify shortfalls in equipment, software, and personnel; and determine the cost to remedy these shortfalls.

As part of this tabletop study, DoD could evaluate its budget and financial management processes with a view toward simplifying and accelerating the operations, modernization, and recapitalization processes. Those evaluations could reveal alternative budgeting and financial management processes, including changes in law and regulation, to enable the ranges to act quickly to provide test and evaluation services for current and future requirements. DoD could also consider the use of a working capital fund specifically for the ranges across DoD. This pilot study could include an evaluation of processes to control costs, such as a rate board made up of customers to evaluate rates charged.

To strengthen the effectiveness of the pilot program, it could:

1. Operate under the general supervision of the Deputy Secretary of Defense.
2. Offer flexibility in funding authorities, such as those stated in the pilot for agile software development (NDAA, 2019).
3. Create a working capital fund to cover operational, recapitalization, modernization and sustainment costs of ranges, with funding mechanisms designed to mitigate program cost-driven incentives to forego testing.
4. Prioritize and correct capabilities gaps in a selected joint technology area with multi-domain test requirements and broad range enterprise implications, such as:
    a. End-to-end operational evaluation of hypersonic weapons, and
    b. Connected concurrent kill chain operations as a capstone OT&E activity.
5. Simplify resource allocation, financial management, acquisition, and any other processes or rules that impede rapid, effective, and efficient funding of ranges and infrastructure, including:
    a. Software-enabled capabilities and the maintenance of software over time, and
    b. Modeling and simulation support.
6. Test the simplified processes using the capabilities gaps identified above.
7. Ensure appropriate cost control through a rate board made up of range customers as part of the new working capital fund mechanism.
8. Include in the pilot project representatives of all affected agencies, who will have decision-making power for their respective agencies.
9. Notify Congress of its findings, conclusions, and recommendations within 18 months of enactment, including:
    a. Identified barriers in policy, regulation, or statute to identification and documenting validated test infrastructure requirements, range modernization, and sustainment of new or orphaned capabilities.
    b. Identified barriers to funding the development, sustainment, and execution of mission-level operational assessments that

focus on multi-system and multi-technology integration for kill chains and joint all-domain operations.

By completing these tasks, DoD will be able to better determine testing requirements expected from the ranges and be better able to fund current and emerging requirements quickly enough to make a difference.

REFERENCES

CJCSI (Chairman of the Joint Chiefs of Staffs Instruction). 2018. "Charter of the Joint Requirements Oversight Council (JROC) and Implementation of the Joint Capabilities Integration and Development System (JCIDS)." Washington, DC. https://www.jcs.mil/Portals/36/Documents/Library/Instructions/CJCSI%205123.01H.pdf?ver=2018-10-26-163922-137.

Conley, K.M., J.R. Dominy, R.R. Kneece, J. Mandelbaum, and S.K. Whitehead. 2014. "Implications of DoD Funds Execution Policy for Acquisition Program Management." Alexandria, VA. https://www.ida.org/~/media/Corporate/Files/Publications/IDA_Documents/SFRD/2014/P-5164.ashx.

CRS (Congressional Research Service). 2020a. *Defense Primer: RDT&E*. Washington, DC. https://fas.org/sgp/crs/natsec/IF10553.pdf.

CRS. 2020b. *Department of Defense Research, Development, Test, and Evaluation (RDT&E): Appropriations Structure*. Washington, DC. https://fas.org/sgp/crs/natsec/R44711.pdf.

Daniels, S.P., and T. Harrison. 2020. "Actual Obligation Rates versus Comptroller Projected Obligation Rates." Center for Strategic & International Studies. Washington, DC. https://csis-website-prod.s3.amazonaws.com/s3fs-public/publication/200302_ObligationRates_v6.pdf?IXrtypFFWHGlMOEzYu94GT.G..bz6ccq.

Department of the Air Force. 2021. *Assessment of the Air Force Test Center: Report to Congressional Committees*. Washington, DC.

GAO (Government Accountability Office). 2016. *Report to Congressional Committees: Defense Facility Condition, Revised Guidance Needed to Improve Oversight of Assessments and Ratings*. Washington, DC. https://www.gao.gov/assets/gao-16-662.pdf.

NDAA (National Defense Authorization Act). 2019. McCain National Defense Authorization Act for Fiscal Year 2019. Public Law 115-232.

6

Conclusion and Summary of Recommendations by Actor

Rapid technological change is driving a new wave of military weapons and technologies, transforming the nature of military conflict itself. Increasingly complex military systems need to be tested and evaluated in ways the Department of Defense (DoD) test ranges had not been originally built to support. Although many DoD test ranges were built during World War II, the ranges have succeeded in advancing many capabilities to enable testing of these emerging technologies. However, military innovation continues evolving faster than the test ranges can keep pace and DoD is at risk of not executing its mission to confirm the operational effectiveness and suitability of defense systems in combat use.

Based on public testimony, site visits to a representative sample of test ranges, test range inputs, and a review of prior unclassified studies, the committee offers a list of the necessary range capabilities highlighted throughout this report that are critical for meeting operational testing needs through 2035 (Box 6.1).

To clarify the obligations of various stakeholders to address these critical needs and ensure the operational superiority of U.S. defense systems through 2035, the following sections parse by stakeholder to the committee's recommendations. While recommendations are assigned to a stakeholder, their implementation will require collaborative efforts by several or all stakeholders listed below.

> **BOX 6.1**
> **Critical Range Capabilities Necessary
> to Test for the Future Fight**
>
> To adequately prepare for the future fight, test ranges will require the following capabilities:
> - High-bandwidth connectivity across ranges, with multi-level security provisions, and common data standards for interoperability (Chapter 4)
> - Overarching, cross-range data strategy, processes, and procedures for collecting, storing, managing, and sharing test data (Chapter 4)
> - Capabilities and success criteria for measuring and evaluating collaboration between systems and end-to-end systems of systems (SoS) performance (Chapter 3)
> - Emulation of physical or threat environments that could affect the closure of the kill chain in an operational setting (Chapter 3)
>
> Among the enabling enterprise needs highlighted in this report are
> - Identification of a process and owner for defining kill chain and multi-domain operation (MDO) doctrine and concepts of operation, which would create cross-program and multi-system test requirements and ultimately drive range capability requirements (Chapter 3)
> - A defined approach to support execution of "beyond program" multi-domain and multiple concurrent kill chain testing (Chapter 3)
> - A defined approach for sustainment of MDO/kill chain joint infrastructure on the ranges, beyond the program that originally built a capability (Chapter 3)
> - On-board data collection systems that capture interactions between systems and actions/decisions driven by interconnected systems for analysis of expected integrated behaviors and outcomes (Chapter 4)
> - Ability to use modeling and simulation (M&S) or live, virtual, and constructive (LVC) approaches to replicate parts of the kill chain or other domains that are impractical in certain test scenarios. This motivates the need for digital infrastructure (Chapter 4)
> - Integrated cross-range "remote" command and control for tests spanning multiple ranges (Chapter 3)

THE RECOMMENDATIONS—BY STAKEHOLDER

Congress should

- Consider mechanisms for increasing the effectiveness and applicability of minor military construction for responsive test and evaluation activities (Conclusion 5-1).

The Department of Defense should

- Establish a joint program effort to enable DoD ranges to test kill chains and joint multi-domain operations (MDOs) that can integrate effects across National Defense Strategy modernization areas (Recommendation 3-1).
- Identify and prioritize bands that cover U.S. military operational and test requirements and preserve these capabilities by protecting them from sell-off, ensuring the ability to validate the survivability of DoD weapon systems against a realistic operational threat environment across air, sea, land, space, and spectrum domains (Recommendation 3-2).
- Broaden the authority of the Test Research Management Center (TRMC) to address issues of internal encroachment by reviewing internal range policies and actions to ensure that the test groups retain adequate mission space and prevent the placement of equipment or infrastructure that could potentially interfere with test operations (Recommendation 3-4).
- Grant the Director of Defense Research and Engineering for Advanced Capabilities the authority to mitigate disputes arising over internal encroachment concerns and provide additional funding to manage internal encroachment (Recommendation 3-4).
- Direct that the Chairman of the Joint Chiefs of Staff require the Joint Requirements Oversight Council (JROC) to consult regularly with the Director of Operational Test and Evaluation, who is an advisor to the JROC, about the test requirements for systems considered by the JROC. This consultation should include an evaluation of current testing capabilities, facilities shortfalls, and plans to address these shortfalls (Recommendation 5-1).
- Either allow an exemption or set shallower expenditure benchmarks for the first 2 years of test modernization programs. This will reflect realistic expense curves for the technologies and projects needed to test next-generation programs and complex integration (Recommendation 5-2).
- Undertake a pilot program that uses a new process and authorities for funding ranges and infrastructure to make them simpler, more responsive, and more effective (Conclusion 5-2).

The Office of the Director of Operational Test & Evaluation should

- Regularly consult with and advise the JROC on the test requirements for systems considered by JROC. This will include an evaluation of current testing capabilities, facilities shortfalls, and plans to address these shortfalls (Recommendation 5-1).

CONCLUSION AND SUMMARY OF RECOMMENDATIONS BY ACTOR

The Test Resource Management Center should

- Assess current and projected commercial radio frequency communications technologies and spectrum allocations for secure, agile, high-bandwidth operational test needs. In addition, TRMC should determine the feasibility of developing new large-scale enclosed testing facilities combined with expanded modeling and simulation to support electromagnetic (EM) spectrum activities not suitable for open-air testing (Recommendation 3-3).
- Address issues of internal encroachment by reviewing internal range policies and actions to ensure that the test groups retain adequate mission space and prevent the placement of equipment or infrastructure that could potentially interfere with test operations (Recommendation 3-4).
- Develop a strategy that assesses the use of and potential investment in suitable allied resources for open-air testing. This strategy should include criteria for the usage of allied resources and areas of potential investment to include range space available, data collection, security risks, and support facilities (Recommendation 3-5).
- Continue monitoring and supporting the Assured Development and Operation of Autonomous Systems (ADAS) Project, and prioritize efforts to develop a common set of standards, measurement approaches, and operational scenarios from which to evaluate the performance of artificial intelligence (AI) and autonomous systems, while recognizing that testing approaches may differ between AI and autonomous systems (Recommendation 4-3).

The Director of Defense Research and Engineering for Advanced Capabilities should:

- Be granted the authority to mitigate disputes arising over internal encroachment concerns and provided additional funding to manage internal encroachment (Recommendation 3-4).

Activities carried out by a Department of Defense joint program effort could include the following:

- Integrate efforts across National Defense Strategy modernization areas to enable DoD ranges to test in a multi-domain battlespace of integrated systems and be capable of testing kill chains and MDOs (Recommendation 3-1).
- Establish a shared, accessible, and secure modeling and simulation (M&S) and data ecosystem to drive development and testing across the life cycles of multiple supporting programs (Recommendation 4-1).

- Adopt and promulgate modern approaches for standardization, architectural design, and security efforts to address data interoperability, sharing, and transmission challenges posed by the complexity of next-generation systems (Recommendation 4-2).
- Determine how to develop and maintain a protected data analysis tool and model repository for testing, increase the interconnectivity of test ranges, and ensure the development of data protocols for the real-time transfer of data at multiple classification levels (Recommendation 4-2).

Appendixes

A

Statement of Task and Completion Matrix

The National Academies of Sciences, Engineering, and Medicine will convene an ad hoc committee to assess the physical and technical suitability of the Department of Defense's (DoD) ranges, infrastructures, and tools used for test and evaluation (T&E) of military systems' operational effectiveness, suitability, survivability, and lethality across all domains (land, sea, air, space, and cyberspace). Specifically, the committee will:

1. Assess the aggregate physical suitability of DoD's ranges to include their testing capacity, the condition of their infrastructure, security measures, and encroachment challenges.
2. Assess the technical suitability of ranges to include spectrum management, instrumentation, cyber and analytics tools, and their modeling and simulation capacity.
3. Evaluate the following attributes for each range:
   - Physical Attributes of Range: Do ranges allow for full exercise of tested systems in the manner that will be used to achieve their mission?
   - Electromagnetic Attributes of Range: Can the system under test, and emulated threats to the system, access and utilize spectrum as designed and needed?
   - Range Infrastructure: Can range instrumentation properly and fully assess system performance and record test data (as well as training data that could be applied to T&E requirements)? Can range tools adequately process and transmit test data and efficiently provide test results?

- Test Infrastructure Security: How secure are ranges, infrastructure, and test capabilities against physical and cyber intrusion that could lead to exploitation of weapon systems performance data by an adversary?
- Encroachment Threats and Impacts: What are the existing and potential future encroachment threats and impacts (physical space, spectrum, alternative/competing DoD uses)?

4. The committee will recommend how DoD can address and/or mitigate any existing or anticipated deficiencies, and test and evaluate future technologies anticipated to arrive between now and 2035, including discussion of planning and resource allocation for the overall test range enterprise. These technologies include, but are not limited to:
    - Directed energy, hypersonic systems, autonomous systems, artificial intelligence, space systems and threats, 6th generation aircraft, advanced acoustic and non-acoustic technologies for undersea warfare, and advanced active electronic warfare/cyber capabilities.

The committee acknowledges the limitations of their data-gathering efforts based on the availability of Distribution A resources to inform their assessments. As a result of these limitations, portions of the statement of task must be addressed by the second phase of this study, which is permitted to utilize resources at higher classification levels than permitted in this phase. Table A.1 is a matrix to illustrate the committee's ability to address the components laid out in its statement of task.

The committee also recognizes the innate challenge of responding to the statement of task language introducing (3): "Evaluate the following attributes for *each* range" (emphasis added). DoD's test and training ranges number over 500 in total, including the 23 major facilities in the MRTFB. DoD does not currently have standardized and comprehensive reporting on test ranges and facilities that would address the items in the statement of task. To assess the current physical and technical state of the test ranges, the committee selected representative ranges spanning all domains (land, sea, air, space, and cyberspace) to provide insights on the aggregate challenges with operational testing unique to each domain. This strategy enables the committee to report on concerns and conditions that were articulated by multiple ranges, services, and agencies. The committee further recognizes that each of DoD's test ranges will face specific challenges and opportunities unique to the individual facility or organization that are not addressed in this report.

APPENDIX A

TABLE A.1 Statement of Task Completion Matrix

		Domains				
		Land	Sea	Air	Space	Cyberspace
Assessment of DoD Test Ranges	Physical Suitability for Testing Capacity • Physical space	**, a	**, a	**, a	*, b	**, a
	Infrastructure Conditions • Buildings • Equipment • Digital infrastructure	**, a	**, a	**, a	*, b	**, a
	Cyber and analytics tools	**, a	**, a	**, a	c	*, b
	M&S capacity	**, a	**, a	**, a	*, c	*, b
	Recommendations to Address Test Deficiencies • Existing • Anticipated	**, a	**, a	**, a	c	**, a
	Range Attributes					
	Security Measures • Physical • Cyber	**, a c	**, a c	**, a c	c c	c c
	Encroachment Challenges • Physical space • Spectrum • Competing missions	**, a **, a **, a	**, a **, a **, a	**, a **, a **, a	*, b *, b *, b	N/A *, b **, a
	Electromagnetic attributes • Spectrum management • Band access and utilization • Threat emulation	**, a *, b c	**, a *, b c	**, a *, b c	*, b c c	**, a *, b c
	Instrumentation • Assess system performance • Record test data • Process and transmit data	**, a	**, a	**, a	c	**, b

Key: ** Fully addressed
 * Partially addressed
 (Blank) Not addressed

a Publicly accessible information enabled committee to address this topic
b Limited to no publicly accessible information available
c Deferred to Phase 2, due to sensitivity of topic

B

Site Visit Summaries

Serving as a significant component of the committee's information-gathering efforts, from March through May 2021, a subset of committee members conducted in-person and virtual informational site visits at six DoD test ranges and received response documents from two additional ranges. The purpose of the site visits was to gather test range perspectives on the statement of task (Box 1.1). The committee selected test ranges that represented different services. They also sought to gather information from both major range and test facility bases (MRTFBs) and non-MRTFBs.

At each in-person or virtual site visit, candid discussions were held with range personnel. As a result of the open and candid discussion held during these site visits, the committee was able to collect unique data, both qualitative and quantitative in nature, on current and projected operational testing and evaluation (OT&E) challenges. The site visit discussions inform several of the findings, conclusions, and recommendations within this report.

Sample site visit questions include the following:

- What are the top three current OT&E challenges facing your range?
- What are the top three future or projected OT&E challenges facing your range?
- What are the encroachment concerns at your range?
- Are there any issues related to OT&E funding sources?
- What is the whole spectrum of your funding stream, including commercial activities, for FY2020?

APPENDIX B   115

Below is a summary of the committee's site visit discussions and response documents:

ABERDEEN PROVING GROUND

Background. Aberdeen Proving Ground (APG) is an Army Major Range and Test Facility Base (MRTFB). Under the command of the Army Test and Evaluation Command (ATEC), the Aberdeen Test Center (ATC) at APG serves as the lead test center for unmanned ground vehicles, vulnerability/lethality, automotive/tracked and wheeled, direct-fire systems, small arms systems, direct-fire weapons performance, and littoral warfare.[1] APG was selected to examine Army Futures activities, including autonomous vehicles, range for vehicle testing, and the roadway simulator. ATEC, in coordination with ATC, provided a response document to a set of questions prepared by the committee to gain insight into the OT&E issues and challenges at APG. Their responses are summarized below.

Current and Future OT&E Challenges. The top three current OT&E challenges outlined by APG were:
1. Securing adequate funding necessary for sustainment, operational, and modernization costs.
2. The ability to expedite data transport, reduction, analysis, and visualization, which requires fiber optic infrastructure and wide area network bandwidth.
3. Personnel skillsets to support modeling and simulation, artificial intelligence, machine learning, and cyber infrastructure.

The top three future OT&E challenges for APG include:
1. The ability to safely operate robotics and autonomous controls.
2. A lack of adaptable test instrumentation, methods, and infrastructure that can be rapidly applied to changing requirements for novel and increasingly complex next-generation weapon systems and accessories.
3. The integration of modeling and simulation with traditional live testing.

Connectivity and Security Challenges. Analysis of secure test data requires the manual transport and processing of classified data from unconnected test events. Connectivity at APG is currently through standard network routers, but there are current investments for fiberoptic

---

[1] CBRNE Central, "Aberdeen Test Center (ATC)," profile, https://cbrnecentral.com/profiles/name/aberdeen-test-center-atc.

modernization. However, that modernization is years away. APG is currently investigating applicability of cellular technologies (4G/5G).

Funding. APG noted that ATC is funded as an Army MRTFB activity for developmental test and evaluation (DT&E). Operational test and evaluation (OT&E) is customer-funded activity and is therefore restricted based on the availability of resources. Current resource restraints on developmental testing (DT) directly affect operational testing (OT) support operations, which are increasing (18 OT support operations in the last year). This restraint makes it difficult to support projects like Cross Functional Teams (CFTs) Acquisition Category (ACAT).

Encroachment. APG noted the following encroachment concerns on operational test activity: climate change, noise, air quality, and spectrum availability.

EGLIN AIR FORCE BASE/EDWARDS AIR FORCE BASE/ AIR FORCE RESEARCH LABORATORY

Background. The Air Force Test Center (AFTC) is overseen by the Air Force Materiel Command (AFMC) and oversees a broad array of test facilities including the 96th test wing at Eglin AFB, the 412th test wing at Edwards AFB, and the Air Force Research Laboratory (AFRL) at Wright-Patterson AFB.[2] The 96th and 412th test wings are Air Force MRTFBs. These sites were selected to better understand the interaction between DT and OT and how connectivity is achieved between ranges that are not in geographical proximity to one another. This visit was conducted virtually on April 6, 2021.

Current and Future OT&E Challenges. The top three current OT&E challenges outlined during the site visit included lack of space to conduct expanding kill chains from launch to target, technical challenges with test infrastructure, and the range capacity with associated scheduling issues. Looking ahead, the representatives viewed the following as future OT&E challenges: (1) Lack of facilities to conduct kill chain testing, including both "on range" and "off range" test infrastructure. The "off range" infrastructure includes facilities for modeling and simulation (M&S), software-in-the-loop (SIL), human-in-the-loop (HITL), and live, virtual, and constructive (LVC) approaches. (2) Cybersecurity testing. The current process is forcing testers to reinvent processes every 16 months.

---

[2] See https://www.edwards.af.mil/News/Article/394391/afftc-re-designated-as-air-force-test-center.

Problem stems from testers not knowing how code is written. (3) Integrating ranges so that larger tests can be conducted. Several representatives also mentioned the need for a national road map for emerging technology areas that includes funding, maintenance, modernization.

Funding. Representatives noted funding shortfalls in a variety of areas including, hypersonic investments, maintenance for legacy systems, and the pot of money available for modernization. Long lead times also results in a slow process in which funding must be advocated for in advance, cannot move fast. Finally, there is not sufficient funding to close identified capability gaps, leading ranges to create patchwork-funding arrangements to close capability gaps.

Encroachment. The range representatives noted that spectrum encroachment and internal encroachment posed the biggest threats to Eglin AFB and Edwards AFB going forward. The representatives provided examples of internal encroachment including the insertion of the 7th Special Operations Forces group at Eglin AFB, the government restricting/taking away telemetry for test, Navy training operations at Fallon and Ramore training in the R2508 airspace, training missions increasingly conducted on test space, and F-16s taking up test space at White Sands Missile Range (WSMR).

MISSILE DEFENSE AGENCY

Background. The Missile Defense Agency (MDA) is a research, development, and acquisition agency within DoD whose mission is to develop and deploy a layered Missile Defense System to defend the United States, its deployed forces, allies, and friends from missile attacks in all phases of flight. MDA was selected to better understand end-to-end testing across multiple ranges. This visit was conducted virtually on March 19, 2021.

Current and Future OT&E Challenges. The top current OT&E challenges noted by MDA included high demand for services, a lack of trained personnel, the inability to share information across ranges, and aging infrastructure and telemetry assets. Looking ahead, the representatives viewed achieving automated flight safety system (AFSS) compliance by 2030 as a future OT&E challenges because AFSS requirements are not set by all programs utilizing MDA's services.

Funding. The current funding model for MDA was recognized by personnel as one that works very well. MDA operates with Integrated Master Test Plans (IMTPs) that are updated twice per year. ".0" informs the program objective memorandum (POM), and ".1" informs the President's budget. The funding requirement includes ranges and flight test rates. Funding includes both research, development, test, and evaluation (RDT&E) for development, operations and maintenance (O&M) for fielded systems, and military construction (MILCON) for both mission and support. The resources provided include both fixed and variable, with the fixed resources generally remaining the same (predictable) in requests and budgets.

Encroachment. MDA noted spectrum encroachment, particularly in S-band, as a primary encroachment concern. Representatives also noted that Alaska Aerospace, formerly Kodiak test site, is facing encroachment because of supplemental environmental assessment restrictions.

NATIONAL CYBER RANGE COMPLEX

Background. The National Cyber Range Complex (NCRC) was created as a Defense Advanced Research Projects Agency (DARPA) function and is now overseen by the Test Resource Management Center (TRMC). The committee was able to engage with NCRC Director AJ Pathmanathan in a virtual setting to discuss the connection between the cyber range and operational test. This visit was conducted virtually on March 24, 2021.

Current and Future OT&E Challenges. NCRC identified four key current OT&E issues: (1) a lack of a trained cyber workforce; (2) a lack of requirements and funding for programs to use a cyber range; (3) compatibility issues are arising from the fact that next-generation systems need to be integrated with older programs, such as Microsoft Windows 98; and (4) an increased need to work with agencies in the intelligence community to close the loop on integrating current threat intelligence into the virtual test environment.

Funding. NCRC noted that cyber testing will become an unfunded mandate if a funding stream for cyber OT is not established. Currently, NCRC tests are paid for by programs that have extra funding available to come to the cyber range for testing.

Encroachment. Encroachment was not identified as a concern during the NCRC site visit.

NEVADA TEST AND TRAINING RANGE

Background. The Nevada Test and Training Range (NTTR) is an Air Force MRTFB. NTTR provides training, test, and developmental testing area and hosts test operations of the 98th Test Wing and other customers for OT&E test. The committee was able to engage in a full-day virtual discussion with individuals from various NTTR elements including plans and programs, operations, financial management, physical security, range support, and program management. This visit was conducted virtually on April 13, 2021.

Current and Future OT&E Challenges. Managing increasing capacity was noted as a major current challenge at NTTR. Additionally, personnel indicated the need to better anticipate future program requirements. A possible solution mentioned was connecting earlier with the DT community to learn about and coordinate test requirements. Another challenge is data transmission speed, with personnel noting that customers may have to wait 50-60 days to receive data results from their test event. Future OT&E challenges included a lack of understanding as to how to incorporate artificial intelligence (AI) into test events, the lack of definitions for MDOs, and the need to purchase radar arrays that are not run on proprietary software. A final future OT&E issue mentioned was that the stove piping of programs, meaning the lack of coordination and integration across programs and missions, makes MDO testing difficult to perform.

Funding. Personnel characterized the funding methods for NTTR as generally good. Funds for future investment in infrastructure and instrumentation are provided by Air Combat Command. Personnel did note that RDT&E funding provides more flexibility and there is a need for more RDT&E funding going forward.

Encroachment. NTTR personnel pointed out multiple areas where encroachment is impacting operations on the range. First, a recent land withdrawal strategy that was rejected resulted in the loss of test space for NTTR.[3] Additionally, NTTR's 2.9 million acres are protected by fencing and require upkeep and audits. Spectrum encroachment also affects the NTTR operations. NTTR no longer receives request for GPS jamming tests because of Federal Aviation Administration (FAA) restrictions. Finally, internal encroachment creates issues for test and training events at NTTR. Foreign partners frequently utilize range space to test aircraft like the F-35 Joint Strike Fighter. The presence of countries such as the United Kingdom, Singapore, Italy, and Australia creates security issues and results in the rescheduling of test events.

---

[3] Proposal to Withdrawal and Reservations of Public Lands in Nevada to Support Military Readiness and Security, https://fas.org/man/eprint/ndaa-2021-prop/04172020-nevada.pdf, accessed June 4, 2021.

ATLANTIC TEST RANGE AND AIR COMBAT ENVIRONMENT TEST & EVALUATION FACILITY AT PATUXENT RIVER

Background. A small subset of the committee was able to travel to the Atlantic Test Range (ATR), a Navy MRTFB, and the Air Combat Environment Test & Evaluation Facility at Patuxent River. The committee sought to better understand integrated physical/virtual testing, as well as operations, financial management, range support, and program management. Of particular interest to the committee was the Joint Simulation Environment (JSE), which provides a high-fidelity modeling and simulation environment to conduct testing on fifth-plus generation aircraft and systems. This visit was conducted in person on March 10, 2021.

Current and Future OT&E Challenges. The current OT&E challenges noted by ATR included (1) the need to have greater test coordination to ensure the most effective use of range time; (2) the lack of a centralized database and repository for M&S and threat references for OT&E; (3) multi-level security and classification raises issues for cross platform testing; and (4) the integration of M&S with OT&E. Projected challenges include the capabilities to test AI systems.

Funding. ATR noted that there were several issues that arose from securing resources for modifying range infrastructure. Representatives from ATR pointed out that the POM process is not versatile enough and that MILCON is not approved fast enough. ATR personnel indicated that they had available resources for supporting range modernization needs, but were constrained because of strict limitations on mixing investment streams.

Encroachment. Spectrum encroachment was noted as a primary concern.

POINT MUGU SEA RANGE

Background. The Point Mugu Sea Range is a Navy MRTFB selected by the committee because of its electronic warfare testing capabilities. This site visit was conducted virtually on April 7, 2021.

Current and Future OT&E Challenges. Personnel from Point Mugu noted that current OT&E issues stem from costly test failures that result from losing or not having enough targets/kill removal systems on the range. These test failures can cost up to $5 million per target lost. The second challenge noted was the struggle to maintain cybersecurity in an ever-changing cyber environment. Future concerns for OT&E include advancements in hypersonic test, multi-level security, infrastructure modernization, keeping pace with advancements in directed energy, lack of understanding for authority on cross-domain testing, and range demand exceeding capacity. A final note by personnel was that many infrastructure upgrades are paid for by program customers. This means that the new infrastructure is locked to the customer that paid for them. Point Mugu therefore can have modern infrastructure but cannot utilize it for any other tests outside the program that paid for the infrastructure.

Funding. Point Mugu personnel stated that there were large unfunded requirements that lead to cascading funding issues. While Point Mugu does receive MRTFB funding, 85 percent of those funds go to personnel costs. The Investment and Modernization (I&M) budget covers high-cost upgrade items, but about a third of these funds go directly toward maintenance on existing infrastructure. Point Mugu personnel stated that the result of these funding issues was that customers bear the brunt of aging infrastructure costs.

Encroachment. Not identified as a major concern.

VANDENBERG SPACE FORCE BASE

Background. Previously the Vandenberg Air Force Base, Vandenberg now supports the United States Space Force (USSF). Vandenberg is a space launch base for USSF, but it also provides space launches for commercial entities and non-defense agencies. The committee selected Vandenberg as a site to understand the complications to future testing for USSF. Vandenberg provided a read-ahead document in response to questions provided by the committee.

Current and Future OT&E Challenges. The current challenges outlined in the response document highlighted how USSF priorities do not always align with DoD T&E. This leads to conflicting priorities for test events. An additional challenge noted was the deconflicting of activities with the Navy sea ranges on the West Coast.

Funding. The response document stated that the current range costs are high and unsustainable. Representatives stated that this is the result of the 2003 National Defense Authorization Act (NDAA) change to the direct cost only model.

Encroachment. Vandenberg personnel shared operational security concerns arising from commercial test, limitations imposed on the electromagnetic spectrum, and the loss of air and sea space for test operations in recent years.

C

Committee Member Biographies

DANA "KEOKI" JACKSON, *Chair*, is senior vice president and general manager, MITRE National Security Sector. In this role, he is responsible for the strategic growth and execution of MITRE's national security programs, including support to the U.S. Department of Defense, the U.S. Department of Justice, and the Intelligence Community. He also leads the National Security Engineering Center. After more than two decades at Lockheed Martin, Dr. Jackson brings robust technical leadership and business experience, including directly contributing to the design, development, deployment, and flight operation of major national security spacecraft and programs. He also held management roles on the GPS III position, navigation, and timing program, and the Space-based Infrared System missile warning program. Dr. Jackson held several executive and senior management roles at Lockheed Martin, including chief technology officer and chief engineer, and vice president of engineering and program operations. He most recently served as vice president of supply chain and program performance and was responsible for program and supply chain management strategy, execution, and success across the enterprise. Before joining Lockheed Martin, Dr. Jackson was a NASA research fellow at the Massachusetts Institute of Technology (MIT) in the field of human adaptation to the space environment. He is a fellow of the United Kingdom Royal Aeronautical Society and the American Institute for Aeronautics and Astronautics (AIAA) and a member of the National Academy of Engineering (NAE), Sigma Xi, the International Academy of Astronautics, and the Institute of Electrical and Electronics Engineers. Dr. Jackson previously served on the Sandia Corporation board of directors, the AIAA

Foundation board of trustees, the Georgia Institute of Technology president's advisory board, the University of Maryland Clark School of Engineering board of visitors, and the MIT Department of Aeronautics and Astronautics visiting committee. He received his bachelor's, master's, and doctoral degrees in aeronautics and astronautics from MIT and completed the Stanford Executive Program at the Stanford Graduate School of Business.

DARRYL AHNER is a professor of operations research at the Air Force Institute of Technology (AFIT). He also serves as the director of the Scientific Test and Analysis Techniques Center of Excellence (STAT COE) at AFIT. Dr. Ahner is a member of the Institute for Operations Research and the Management Sciences (INFORMS), the International Test and Evaluation Association (ITEA), and the Military Operations Research Society (MORS). Dr. Ahner graduated from the United States Military Academy at West Point, earned an M.S. in applied mathematics, an M.S. in operations research and statistics, and his Ph.D. in systems engineering while serving as a Charles Stark Draper laboratory fellow. Dr. Ahner has earned awards including the 2019 Air Education and Training Command Analysis Team of the Year and the Wilbur B. Payne Award for the best Army analytical technical study.

KAREN BUTLER-PURRY is the associate provost for graduate and professional studies at Texas A&M University where she also serves as a professor of electrical and computer engineering, and as the assistant director of the Power System Automation Laboratory. Dr. Butler-Purry conducts research at Texas A&M on protection and control of distribution systems and isolated power systems such as all electric power systems for ships, mobile grids, and microgrids, cybersecurity protection, intelligent systems for equipment deterioration and fault diagnosis, and engineering education. Prior to joining Texas A&M, Dr. Butler-Purry held technical positions at Hughes Aircraft Company Radar Systems Group, IBM, and MIT Lincoln Laboratory. She is a fellow at the Institute of Electrical and Electronics Engineers (IEEE). She holds her B.S. in electrical engineering from Southern University at Baton Rouge, an M.S. in electrical engineering from The University of Texas at Austin, and her Ph.D. in electrical engineering from Howard University.

GRAHAM CANDLER is the Russell J. Penrose and McKnight Presidential Chair in Aerospace Engineering and Mechanics at the University of Minnesota. He uses computational methods to study high-speed flight with application to future hypersonic flight systems and the entry of spacecraft into planetary atmospheres. He is recognized by the National

Academy of Engineering for development and validation of computational models for high-fidelity simulation of supersonic and hypersonic interactions. Candler and his research collaborators have developed widely used computational methods and codes that are being used for the design and analysis of future hypersonic flight systems, including several NASA exploration missions. Recently, his work has focused on the development of high accuracy simulation methods for the exploration of hypersonic flight system design space. He has published extensively in the areas computational methods, high-temperature gas dynamics, boundary layer laminar to turbulent transition, and validation of computational simulations with hypersonic wind tunnel data. Candler has been at the University of Minnesota since 1992, and leads a research group in hypersonic aerodynamics and computational fluid dynamics. He has received numerous awards, including the American Institute of Aeronautics and Astronautics (AIAA) Thermophysics Award (2007) and Fluid Dynamics Award (2012). He is a Fellow of the AIAA. Candler received his undergraduate degree from McGill University in 1984 and his graduate degrees in aeronautics and astronautics from Stanford University in 1985 and 1988.

GORDON FORNELL served in the USAF for 35 years, retiring in 1993 as a Lieutenant General. He flew 200 combat missions in Vietnam War in the A-1 Skyraider, served as a C-5A operational commander, KC-10 program director, test pilot, and held senior acquisition leadership positions. He was the senior military assistant to both Secretaries of Defense Casper Weinberger and Frank Carlucci. Lt. Gen. Fornell holds a B.S. in mechanical engineering from Michigan State University and an MBA from the Wharton School, University of Pennsylvania.

DERRICK HINTON serves as the vice president for Research and Engineering Services for the Scientific Research Corporation (SRC) in the Advanced Technology Solutions Division. Prior to joining SRC in 2018, Mr. Hinton was a member of the Senior Executive Service with a 25-year civilian career in the Department of Defense (DoD). In his most recent role as acting director, Test Resource Management Center (TRMC), Mr. Hinton advised the Secretary of Defense and the Under Secretary of Defense for Acquisition, Technology and Logistics [USD(AT&L)] on all matters pertaining to the DoD Major Range and Test Facility Base (MRTFB), the nation's critical range infrastructure for conducting effective test and evaluation (T&E). In addition, Derrick oversaw the management of the Central Test and Evaluation Investment Program (CTEIP), the Test and Evaluation/Science and Technology (T&E/S&T) Program, and the Joint Mission Environment Test Capability (JMETC) Program, whose annual budgets collectively totaled over $300M. He also oversaw the management of the National Cyber Range Complex (NCRC) and served as the DoD executive agent for Cyber Test Ranges. Derrick began his career serving in the United States Marine Corps Reserve from 1985 to 1991, and entered the DoD civilian workforce in 1989 as a test engineer responsible for munitions T&E with the 46th Test Wing at Eglin Air Force Base, FL. In 1996, Mr. Hinton joined the AT&L team, initially serving in the Office of the Director, Test, Systems Engineering, and Evaluation. He transitioned to the Office of the Director, Operational Test and Evaluation (DOT&E) in 2001 and joined the TRMC in 2005, taking on the role of principal deputy director, TRMC in 2009. Mr. Hinton holds a bachelor's degree in industrial engineering from the University of Alabama, and a master's of public administration, and an Acquisition Core Level III Certification in Test and Evaluation from the Defense Acquisition University.

ROB KEWLEY currently serves as a director and systems engineer at simlytics.cloud LLC. Prior to that, he served as the acting executive director of the Office of the Chief Systems Engineer. In this position, Dr. Kewley was responsible for developing systems engineering capabilities and processes for Army modernization. At West Point, Dr. Kewley served as head of the Department of Systems Engineering and the United States Military Academy director of operations research. In this position, he led studies in support of Army and DoD analytic challenges. Dr. Kewley received his B.S. in mathematics from West Point, and an M.S. in industrial and management engineering and a Ph.D. in decision science and engineering systems both from Rensselaer Polytechnic Institute.

LAURA McGILL (NAE) is currently the Deputy Laboratories Director and Chief Technology Officer for Nuclear Deterrence at Sandia National Laboratories. Prior to this role, she served as the deputy vice president of engineering at Raytheon's Missiles & Defense, a subdivision of Raytheon Technologies Corporation. Previously, she served as the vice president of engineering at Raytheon Missile Systems. Ms. McGill also served as the product line chief engineer for air warfare systems. Ms. McGill served as an adjunct lecturer for Raytheon's onsite M.S. in systems engineering program in conjunction with Johns Hopkins Whiting School of Engineering. She is a Lifetime Fellow of the American Institute of Aeronautics and Astronautics (AIAA). Ms. McGill was elected to the National Academy of Engineering in 2019. She earned her bachelor's degree in aerospace, aeronautical and astronautical engineering from the University of Washington. McGill also holds a master's degree in aerospace systems from West Coast University.

HANS MILLER is a system test engineer and project leader at the MITRE Corporation. Prior to that, he was the division chief of policy, programs and resources at the USAF Headquarters for Test and Evaluation. He has 25 years of experience in the Air Force as a test pilot, program manager, and commander of large flight and ground test organizations. Mr. Miller also has experience working with the international partners though a NATO assignment and as the program manager of the DoD Foreign Comparative Test Program. Mr. Miller graduated from the United States Air Force Academy with a bachelor's degree in aeronautical engineering and a master's of aeronautics and astronautics from Stanford University. He also attended the USAF Air War College, USAF Test Pilot School, and USAF Weapons School.

HEIDI C. PERRY is currently assistant division head for the Air, Missile and Maritime Defense Technology Division at the Massachusetts Institute of Technology Lincoln Laboratory. In her role, she works strategic initiatives for undersea systems and serves as the chief innovation officer for the division. Previously, Ms. Perry was director, system engineering, at the Charles S. Draper Laboratory, Incorporated. She also served in other senior leadership roles, including director, algorithms & software, and director, internal R&D portfolio. Her expertise includes guidance, navigation, and control; global position system anti-jam and ground control; autonomous systems; mission-critical software; and command, control, communications, computers, intelligence, surveillance, and reconnaissance systems. Ms. Perry began her career with General Electric as a systems engineer working on the AN/BSY-2 Sonar System before moving to IBM, as a systems engineer for avionics design and flight test programs. From IBM she moved to Draper Laboratory as task leader for the Dolphin Navigation System Upgrade and remained with Draper for more than 20 years. In these years at the laboratory, she served as technical director for various research and development programs involving autonomous spacecraft, aircraft, robotics systems, and underwater vehicles. A former member of the Naval Studies Board (2008–2013), she also served on the National Academies' Committee on Capability Surprise for U.S. Naval Forces, Committee on National Security Implications of Climate Change on U.S. Naval Forces, and Committee on the "1,000 Ship Navy"—A Distributed and Global Maritime Network. She served as the co-chair of the National Academies' Committee on Mainstreaming Unmanned Undersea Vehicles into Future U.S. Naval Operations and recently served as the chair for the Transportation Research Board's Committee on Leveraging Unmanned Systems for Coast Guard Missions. She received a B.S. in electrical engineering from Cornell University and an M.S. in computer engineering from the National Technical University. She currently serves as a member of the President's Council of Cornell Women.

GARY F. POLANSKY is a senior scientist at Sandia National Laboratories and has worked for more than 35 years in national security, nuclear energy, and environmental programs. His broad-based technical capabilities have made key contributions to many program areas, including aerospace systems, space nuclear power and propulsion, nuclear energy, and nuclear materials management. He currently has wide ranging technical responsibilities across programs in the Integrated Military Systems Development Center. Dr. Polansky was the program manager for the highly successful Advanced Hypersonic Weapon Flight Test 1A. This test demonstrated a first-of-its-kind vehicle that was designed to fly through the atmosphere at hypersonic speed and long range. The flight test team was recognized with a Lockheed Martin Nova Award. Dr. Polansky has authored or co-authored more than 50 technical publications in computational physics, nuclear technology, and hypersonic systems. He has served as session chair and conference organizer for national and international technical conferences. He is a Fellow of both the American Society of Mechanical Engineers (ASME) and the American Institute of Aeronautics and Astronautics (AIAA). He has served on national technical committees and was active in local professional society activities for many years. He received a Ph.D. in engineering from The University of Texas at Austin.

KARL F. SCHNEIDER previously served as the senior official to perform the duties of the Under Secretary of the United States Army. The Under Secretary of the Army performs the duties of the Secretary of the Army's senior civilian assistant and principal adviser on matters related to the management and operation of the Army, including development and integration of the Army program and budget. Prior to this position, Mr. Schneider served as the Principal Deputy, Assistant Secretary of the Army (Manpower and Reserve Affairs), providing oversight of all planning, analysis and assessment support to the Total Force manpower and personnel policy. He served in that position from September 20, 2013, until April 18, 2014.

WILLIAM WILSON is the acting director of the CERT Division at the Software Engineering Institute at Carnegie Mellon University. In this position, he works to identify new technologies, system development practices, and management practices to improve network systems. Previously, Mr. Wilson served as the technical manager of CERT's Survivable Enterprise Management Initiative where he developed enterprise security management and information security risk assessment methods. Before joining Carnegie Mellon, Mr. Wilson worked as the technical director of the Engineering Center at the National Security Agency, where he served for 12 years. Mr. Wilson holds a bachelor's in computer science from Pennsylvania State University and a master's degree in computer systems management from the University of Maryland.

D

Disclosure of Unavoidable Conflicts of Interest

The conflict-of-interest policy of the National Academies of Sciences, Engineering, and Medicine (https://www.nationalacademies.org/about/institutional-policies-and-procedures/conflict-of-interest-policies-and-procedures) prohibits the appointment of an individual to a committee like the one that authored this Consensus Study Report if the individual has a conflict of interest that is relevant to the task to be performed. An exception to this prohibition is permitted only if the National Academies determine that the conflict is unavoidable and the conflict is promptly and publicly disclosed.

When the committee that authored this report was established a determination of whether there was a conflict of interest was made for each committee member given the individual's circumstances and the task being undertaken by the committee. A determination that an individual has a conflict of interest is not an assessment of that individual's actual behavior or character or ability to act objectively despite the conflicting interest.

Dr. Keoki Jackson was determined to have a conflict of interest because of his prior affiliation with Lockheed Martin, which develops products for the Department of Defense (DoD), many of which undergo operational test and evaluation at DoD ranges that are included in the study. Lockheed Martin also has an operations contract with the National Cyber Range Complex, which is under the purview of DoD's Test Resource Management Center and the location of one of the site visits for this study.

Mr. Derrick Hinton was determined to have a conflict of interest because of his current affiliation as an employee of the company Scientific

Research Corporation (SRC), whose business activities are focused on a broad range of information, communications, intelligence, electronic warfare, simulation, training, and instrumentation systems for both commercial and defense operational environments.

Dr. Rob Kewley was determined to have a conflict of interest because of his current affiliation as a consultant for multiple companies that compete for modeling and simulation support for DoD programs, including programs in the test domain.

In each case, the National Academies determined that the experiences and expertise of these individuals were needed for the committee to accomplish the task for which it was established. The National Academies could not find another available individual with the equivalent experiences and expertise who did not have a conflict of interest. Therefore, the National Academies concluded that the conflict was unavoidable and publicly disclosed it on its website (www.nationalacademies.org).

E

Abbreviations and Acronyms

ABMS	Advanced Battle Management System
ACAT	Acquisition Category
ACETEF	Air Combat Environment Test and Evaluation Facility
ADAS	Assured Development and Operation of Autonomous Systems
AFB	Air Force Base
AFOTEC	Air Force Operational Test and Evaluation Center
AFRL	Air Force Research Laboratory
AFSS	Automated Flight Safety System
AFTC	Air Force Test Center
AI	artificial intelligence
APG	Aberdeen Proving Ground
ARL	Army Research Laboratory
ATEC	Army Test and Evaluation Command
ATR	Atlantic Test Range
BOARD	Board on Army Research and Development
C2	command and control
CEC	cooperative engagement capability
CFT	Cross Functional Teams
CI/CD	continuous integration/continuous delivery
CNO	Chief of Naval Operations
COCOM	Combatant Command
COMOPTEVFOR	Commander, Operational Test and Evaluation Force

CPP	Conservation Partnering Program
CPS	cyber-physical systems
CRIIS	Common Range Integrated Instrumentation System
CRS	Congressional Research Service
CTEIP	Central Test and Evaluation Investment Program
DARPA	Defense Advanced Research Projects Agency
DAU	Defense Acquisition University
DBCRC	Defense Base Closure and Realignment Commission
DNWR	Desert National Wildlife Range
DoD	(U.S.) Department of Defense
DOE	(U.S.) Department of Energy
DOT&E	Director of Operational Test and Evaluation
DT	developmental testing
DT&E	developmental test and evaluation
EA	electronic attack
EGTTR	Eastern Gulf Test and Training Range
EM	electromagnetic
EO/IR	electro-optical and infrared
EW	electronic warfare
EWS	electronic warfare support
F2T2	find, fix, track, target
F2T2EA	find, fix, track, target, engage, assess
GAO	Government Accountability Office
GPS	Global Positioning System
HITL	human-in-the-loop
I&M	Investment and Modernization
IMTP	Integrated Master Test Plan
IT	information technology
JADO	Joint All-Domain Operations
JAIC	Joint Artificial Intelligence Center
JCIDS	Joint Capabilities Integration and Development Systems
JIM	Joint Improvement and Modernization
JMETC	Joint Mission Environment Test Capability
JROC	Joint Requirements Oversight Council
JSE	Joint Simulation Environment
JSF	Joint Strike Fighter

APPENDIX E

JT&E	Joint Test and Evaluation	
LVC	live, virtual, and constructive	
M&S	modeling and simulation	
MBSE	model-based systems engineering	
MDA	Missile Defense Agency	
MDO	multi-domain operation	
MILCON	Military Construction	
MRTFB	Major Range and Test Facility Base	
NASEM	National Academies of Sciences, Engineering, and Medicine	
NCRC	National Cyber Range Complex	
NDAA	National Defense Authorization Act	
NRC	National Research Council	
NSF	National Science Foundation	
NSTTR	National Space Test and Training Range	
NTTR	Nevada Test and Training Range	
O&M	operations and maintenance	
ONR	Office of Naval Research	
OODA	observe-orient-decide-act	
OSD	Office of the Secretary of Defense	
OT	operational testing	
OT&E	Operational Test and Evaluation	
POM	program objective memorandum	
R&D	Research and Development	
R&E	Research and Engineering	
RDT&E	research, development, test, and evaluation	
REPI	Readiness and Environmental Protection Initiative	
SDS	spectrum dependent systems	
SHADE	Shared Data Environment	
SIL	software-in-the-loop	
SoS	system-of-systems	
SRI	Sustainable Ranges Initiative	
T&E	test and evaluation	
T&E/S&T	Test and Evaluation/Science and Technology	
TEL	Transporter Erector Launcher	

TEMP	Test and Evaluation Master Plan
TRMC	Test Resource Management Center
USSF	United States Space Force
WSMR	White Sands Missile Range